Minerals, Rocks and Inorganic Materials

Monograph Series of Theoretical and Experimental Studies

7

Edited by

W. von Engelhardt, Tübingen · T. Hahn, Aachen
R. Roy, University Park, Pa. · P. J. Wyllie, Chicago, Ill.

A. Rittmann

Stable Mineral Assemblages
of Igneous Rocks

A Method of Calculation

With Contributions by
V. Gottini · W. Hewers · H. Pichler · R. Stengelin

With 85 Figures

Springer-Verlag New York · Heidelberg · Berlin 1973

Professor Dr. Dr. h. c. *Alfred Rittmann*
Istituto Internazionale di Vulcanologia
I-95 123 Catania

ISBN 0-387-06030-8 Springer-Verlag New York · Heidelberg · Berlin
ISBN 3-540-06030-8 Springer-Verlag Berlin · Heidelberg · New York

Preface

This book represents the results of a lengthy study which Professor ALFRED RITTMANN began some thirty years ago. The relationship between the chemical and mineralogical composition of igneous rocks is established as far as is possible. Petrographers will appreciate that this problem is extremely complex, particularly since this relationship forms the basis of the classification and nomenclature of igneous rocks.

The ingenious scheme of calculation of the C.I.P.W. norm system is essentially chemical in nature. The compositions of the stoichiometrically ideal "normative minerals" do not correspond to those of the constituent minerals found in rocks. Although the "norm" is not intended to equal the "mode" or actual mineral composition of a rock, at least a qualitative agreement between the norm and the mode is desirable. For a number of rocks and rock groups, especially the leucocratic and silicic rocks, the deviation of the norm from the mode is generally within tolerable limits. For the melanocratic and highly subsilicic rocks, on the other hand, the C.I.P.W. scheme of calculation too often yields results which fail to reflect the observed mineral composition. The anomalies produced in the calculation of extremely subsilicic volcanic rocks have recently been briefly discussed by F. CHAYES and H. S. YODER, JR. (1971).

The frequent disagreement of the CIPW norm with the mode inspired Professor RITTMANN to develop a new method of norm calculation. The new scheme was originally applied to volcanic rocks only, but was later modified and extended for use with subvolcanic and plutonic rocks. To distinguish the result of the new calculation method from those of the C.I.P.W. system the new norms are here called the *Rittmann norms*. The intention of the authors is not to suggest that the *Rittmann norm* should replace the C.I.P.W. norm, but merely to supply an alternative method of calculation. It is up to the reader to decide which one of the two norm calculation schedules is better adapted to this purpose.

Professor RITTMANN approaches the problem in an entirely empirical manner. The average compositions of the main constituents in a certain rock group are estimated on the basis of the available analytical data. The averages found are used in calculating the normative composition of that rock group. A simple example is offered by the silica content of the mineral nepheline. In moderately subsilicic phonolites nepheline generally contains an excess of silica, whereas virtually no silica excess occurs in the nepheline of strongly subsilicic nephelinites and melilite nephelinites. RITTMANN's scheme of calculation is adjusted to take into account this variation in the silica content of nepheline.

The empirical approach is further illustrated by the method of calculating normative nepheline and leucite. The valuation of a number of feldspathoidal rock analyses reveals that the modal occurrence of these two feldspathoids, alone or together, depends not only on the ratio $K/(K + Na)$ in the bulk rock but also on the fluidity of the magma. The salic magmas are more viscous than the femic ones and, accordingly, are more apt to retain their gases. The stability field of leucite is thus reduced. Following this principle, the calculation scheme proceeds separately for the fluid and for the viscous magmas. The "fluid" and the "viscous" magmas are numerically distinguished from each other in the calculation. The procedure gives not only a qualitative but also a semi-quantitative agreement between the norm and the mode.

It must be emphasized that the Rittmann norms are just norms, although they are somewhat more "realistic" than the C.I.P.W. norms. The normative composition which results from the new calculation scheme merely *approaches* the mode and indicates only the *proportions* of the various constituents, not their precise composition. The Rittmann norm is intended to represent an approximation of the *Stable Mineral Assemblage* which would result if the complete crystallization of the magma proceeded slowly enough to maintain chemical equilibrium. Complete agreement, both qualitative and quantitative, between the Rittmann norm and the mode cannot be expected for every single rock. Probable causes of some discrepancies are discussed in a separate chapter. A detailed comparative review of the C.I.P.W. norm and the Rittmann norm is made in the contribution by VIOLETTA GOTTINI.

The book contains detailed *Key Tables* to assist the reader in calculating rock analyses. The scheme looks somewhat lengthy and necessarily involves a number of steps. However, it should be borne in mind that for a given rock only a fraction of the steps need actually be used. The application of the scheme to a number of rock analyses selected as examples will guide the reader in performing his own calculation. The examples demonstrate how disagreements can be eliminated and how they are to be interpreted.

For the convenience of the reader who likes to include a large number of analyses, the schedule has been computerized by R. STENGELIN and W. HEWERS. The program written in ALGOL for a CD 3300 computer is produced as a separate booklet[1]. To enable the scheme to be com-

---

1 Available, price $ 7.00, from Dr. R. STENGELIN, Mineralogisch-Petrographisches Institut der Universität, D-74 Tübingen, Wilhelmstr. 56, W. Germany.

puterized, the various relationships appearing in the Key Tables are given in the form of numerical equations.

The use of the Rittmann norm calculation scheme is illustrated in the contribution by H. PICHLER and R. STENGELIN.

The norm system here presented is intended mainly for the clarification of igneous rock nomenclature, about which some confusion admittedly exists in the literature. It is hoped that the book will make a contribution towards a more uniform nomenclature.

Helsinki, April 1973   TH. G. SAHAMA

Acknowledgments

As long ago as 1941 the author attempted to devise a system of graphs which would yield an approximation of the relation between the mode and the chemical composition of volcanic rocks. At the request of G. B. ESCHER, then president of the International Association of Volcanology (I. A. V.), a more accurate system of graphs was published in 1952 to facilitate the interpretation of chemical rock analyses by volcanologists not specialized in petrography. This preliminary solution of the problem was not intended for use by petrographers who needed a more accurate system of calculating.

In the meantime, a wealth of new data has become available and this permitted a more extended treatment on a statistical basis, which, however, soon showed that the solution of the problem was much more complex than had initially appeared. The results of experimental petrology helped to clarify the process of crystallization of magmas, but in many cases these results could not be extrapolated to the much more complicated natural systems. The further elaboration of the system had to be based on empirical data.

At the request of the mining division of the C. N. E. N. (Comitato Nazionale dell'Energia Nucleare, Rome), the author presented in 1967 a preliminary report entitled "Calcolo delle Associazioni Mineralogiche Stabili nelle Rocce Ignee" which has been used extensively by M. MITTEMPERGHER and colleagues. The author is indebted to him and to the members of his research team for useful discussions and for the supply of new data.

During the further elaboration of the calculation method a series of problems arose regarding phase equilibria and heteromorphic relations of certain rocks about which the author had little personal experience. Many problems had to be discussed with experienced petrographers. The author is particularly grateful to G. MARINELLI, Pisa, G. A. MACDONALD, Oahu, Hawaii, M. MITTEMPERGHER, Rome, TH. G. SAHAMA, Helsinki, and A. STRECKEISEN, Bern, for critical discussions and useful suggestions.

In order to refine the calculation, it was necessary to apply it to the greatest possible number of analyses. This was possible through the great help of H. PICHLER and R. STENGELIN, Tübingen, who programmed the new norm system for the computer and calculated thousands of analyses. This permitted a continual feedback and successive improvement of the method. The author is deeply indebted to these efficient collaborators for their patient assistance and support. Thanks are due also to the author's collaborators V. GOTTINI, R. ROMANO and L. VILLARI, Catania, for valuable assistance. He owes particular thanks to TH. G. SAHAMA and G. A. MACDONALD for having reviewed the manuscript and to H. PICHLER and R. STENGELIN for editorial assistance.

Catania, April 1973   A. RITTMANN

Contents

I. Introduction

1. Previous Methods of Calculation

It is a tempting idea to derive the quantitative mineral assemblage of an igneous rock from the chemical bulk composition. One such attempt was made by W. CROSS, J. P. IDDINGS, L. V. PIRSSON, and H. S. WASHINGTON in 1903 and is known as the C.I.P.W. norm system. In this system the result of the rock analysis is converted to a standard set of normative compounds using a fixed calculation scheme. The contents of the selected compounds, taken as stoichiometrically ideal compounds, are expressed in weight percent. The scheme was essentially based on observations made on volcanic rocks of the calc-alkalic and sodic series.

In fact, the result obtained for the normative mineral composition agrees in many cases tolerably well with the modal or actually observed mineral composition. More often, however, no parallelism is found between the norm and the mode. The rigid scheme of the norm calculation is of a "pigeon hole" type and cannot be adjusted to suit the actual compositions of the minerals found in the rock.

The molecular norm calculation proposed by P. NIGGLI in 1936 is more elastic. Despite also being based on stoichiometrically ideal compounds, this method results in a better approximation to the mode. Compared with the C.I.P.W. system, the advantage encountered in the Niggli norm is the possibility of calculating not only the so-called "Kata-Norm" which, as femic silicates, includes not only pyroxenes and olivine, but also other constituents of the mode, i.e. biotite, amphiboles and melilites. In most cases such a calculation permits, within certain limits, a mutual control between the mode and the bulk chemical composition of the rock.

In his well-known compilation, W. E. TRÖGER (1935, 1939) presented the mineral composition of about a thousand prototypes of igneous rocks. Some of these modes have been measured with the Rosiwal method, others have been only estimated, but most of them have been calculated from the chemical analyses according to an unpublished series of equations[1]. These equations were based on estimated average compositions of the principal constituents in a number of rock types.

---

1 Personal communication of the late W. E. TRÖGER with the author in 1960.

The particular minerals used for deriving the equations are listed on pp. 343–346 in W. E. TRÖGER (1935). In such a calculation the number of equations is usually greater than the number of constituents to be determined. Such an over-determined system makes it possible to choose among the various equations in such a way as to yield a mineral assemblage most closely in agreement with the mode observed under the microscope. In consideration of the great variability of the composition of pyroxenes and amphiboles, this kind of calculation can yield only approximate data which, in addition, are affected by the personal judgement of the petrographer. Furthermore it must be stated that a great number of the rocks mentioned, particularly the plutonic ones, are more or less altered as shown by the presence of secondary minerals and, chemically, by relatively very high contents of water, ferric iron and carbon dioxide or by an excess of alumina over the molecular sum of alkalis and lime. A certain number of chemical analyses are surely erroneous, including not only those dating from the last century. For these reasons a great number of analyses reproduced by TRÖGER (1935, 1939) have not been considered in this book.

2. Nomenclature of Volcanic Rocks

Most petrologists agree with the idea that igneous rocks must be classified according to their actual mineral composition expressed in volume percent, i.e. according to their mode. Extending this principle to volcanic rocks, the naming and classification would be applicable only to holocrystalline rock varieties, the modes of which can be determined under the microscope by point counting or in some other way. Unfortunately, only the basic volcanic rocks frequently exhibit a holocrystalline texture. The more acidic ones usually are partly or mainly glassy. To assign correct names to the semicrystalline or glassy rocks the mineral composition must be calculated from the chemical bulk analysis.

It is a regrettable fact that in the denomination of the volcanic rocks some confusion exists in the literature. A discrepancy between the name of an effusive rock and its chemical analysis is not uncommon. The naming of a volcanic rock is often based merely on the phenocrysts without taking into account the microcrystalline or glassy groundmass. The chemical bulk analysis reflects the composition of the entire rock and indicates the type of magma to which the rock belongs. Therefore, the denomination of a volcanic rock merely on the basis of the optical determination of its phenocrysts is insufficient. The fact that many petrographers and geologists are guilty of this omission may lead to serious consequences in drawing geological, magmatological and geophysical conclusions.

It is recommended, therefore, that the proposal by P. NIGGLI (1931) be followed, i.e. to add to the name of a semicrystalline rock the prefix "pheno" in cases where the rock has been named only on the basis of its crystalline constituents determined under the microscope without chemical bulk analysis. For example, the designation pheno-andesite would mean that the rock has been studied only under the microscope (or macroscopically) and that its true composition might not correspond to that of a true andesite.

If a chemical analysis is available, it is recommended not to base the rock name on microscopic study, but on an accurate calculation of the analysis. In cases where the designation given to the rock is inconsistent with that calculated from the analysis, the name used must be a pheno-name. In order to indicate the microscopical aspect of a

Table 1. Difference in nomenclature of volcanic rocks between literature and new denomination according to the Rittmann norm. Examples: 633 analyses of young volcanics of Japan (from K. ONO, 1962) and 232 of the Hawaiian islands (from G. A. MACDONALD and T. KATSURA, 1964, and others). Analyses of glassy volcanic rocks were not considered. After R. STENGELIN (unpublished)

Rock type	Japan		Hawaii	
	Number of analyses according to literature	Number of analyses according to Rittmann norm	Number of analyses according to literature	Number of analyses according to Rittmann norm
Alkali rhyolites	—	3	—	—
Rhyolites	35	3	—	—
Rhyodacites	1	27	1	1
Dacites	54	191	—	—
Plagidacites	—	108	—	—
Alkali trachytes	—	1	2	4
Trachytes	—	—	4	—
Latites	—	4	—	—
Latiandesites	38	58	—	—
Andesites	406	218	2	—
Latibasalts	—	—	—	1
Mugearites	—	2	3	8
Hawaiites	—	16	16	21
Tholeiitic basalts	—	2	104	110
Basalts ⎱	99	—	36	26
Olivine basalts ⎰		—	39	24
Alkali basalts	—	—	13	11
Phonotephrites	—	—	—	18
Tephrites	—	—	12	8

rock, one can combine its real name deduced from the chemical analysis using the pheno-name as an adjective to the real name, for example: pheno-andesitic dacite.

Table 2a. Difference in nomenclature between literature and new denomination according to the Rittmann norm of 181 analyses of young volcanic rocks of the south-Aegean region, Greece (from H. PICHLER and R. STENGELIN, 1968, p. 800)

Old rock name according to literature	Number of analyses according to literature	Number of analyses according to Rittmann norm	New rock name (according to the Streckeisen classification)
Alkali rhyolites	6	21	Alkali rhyolites
Rhyolites	24	17	Rhyolites
Rhyodacites	2	14	Rhyodacites
Dacites and Dacitoides	52	29	Dacites
—	—	5	Plagidacites
Alkali trachytes	1	—	Alkali trachytes
Trachytes	4	3	Trachytes
Trachydacites	4	—	—
—	—	11	Latites
Trachyandesites	9	57	Latiandesites
Dacitoandesites	5	—	—
Andesites	54	19	Andesites
Andesitic basalts	1	—	—
Basalts	2	—	Basalts
Olivine basalts	1	—	—
—	—	1	Phonotephrites
Without name	17	—	—
Total	181	181	

Table 2b. Distribution (in percent) of 181 young volcanic rocks of the south-Aegean region arranged according to their frequency. The percentage of the old names are given in brackets

Amount in percent		Rock names
31.5	(2.2)	Latiandesites
16.0	(28.7)	Dacites
11.4	(2.8)	Alkali rhyolites
10.5	(29.8)	Andesites
9.4	(13.2)	Rhyolites
7.8	(1.1)	Rhyodacites
6.1	(—)	Latites
3.9	(2.2)	Trachytes
2.8	(—)	Plagidacites
0.6	(—)	Phonotephrites

In order to exemplify the frequency of such discrepancies, in the Tables 1 and 2 the names given by the authors on the basis of microscopic observations (left columns) are confronted with those resulting by calculating the potential modes from the chemical bulk analyses. According to these tables most so-called andesites are pheno-andesitic dacites or pheno-andesitic latiandesites. Plotting the potential modes of these rocks into the Streckeisen double-triangle (Fig. 3) one obtains their distribution areas as illustrated in Figs. 1 and 2.

The necessity of using the prefix "pheno" is further demonstrated by the fact that the specimens collected from a chemically homogeneous

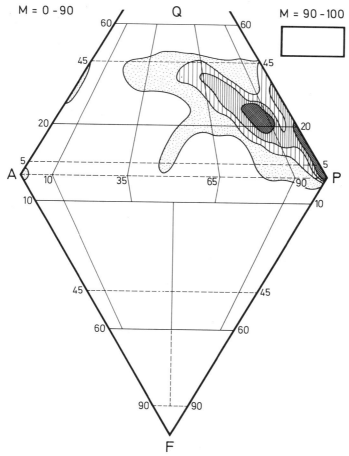

Fig. 1. Distribution areas of 633 young volcanic rocks of Japan. [After R. STEN-GELIN (unpublished); analyses from K. ONO, 1962]

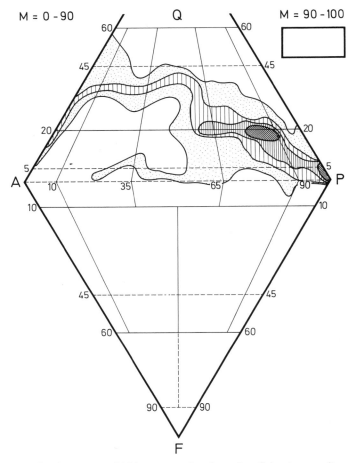

Fig. 2. Distribution areas of 181 young volcanic rocks of the Aegean Sea region/
Greece. (After H. PICHLER and R. STENGELIN, personal communication)

lava flow may be named differently according to their degree of
crystallization. So, for instance, a specimen of mugearite containing
about 60 % glass and phenocrysts of labradorite (20 %), augite and
some grains of olivine, would be designated as a "basalt" by many
authors. A more crystalline specimen of the same flow will appear
as an "andesite", the andesine plagioclase being greatly predominant
over the femic minerals. In a still more crystalline specimen, sanidine
will also have formed in sufficient amount to justify the name "latiandesite". However, there still remains a doubt as to the real com-

Table 3. Examples of different rock type nomenclature

Volcanic rock name according to the Streckeisen system	Local name	NIGGLI's Magma type	C.I.P.W.
Leucite basanite	Kivite	theralite-gabbroid-lamprosommaitic	II (III), 6, (2), 3, 3″
Olivine trachyte	Ciminite	lamprosyenitic	II″, 5, 2, 2
Mela olivine leucitite	Batukite	pyroxenolitic	IV, 1 (2), 2, 2 (3), 1 (2)
Phonotephrite	Tautirite	nosycombitic	I (II), (5) 6, 2, 3

position of the 15–20 % of glass which fills the interstices. Whether it contains potential quartz or nepheline and what the content may be in TiO_2 will determine the definite name of the rock. Only in a relatively coarse-grained holocrystalline specimen can all sanidine and nepheline be detected under the microscope, revealing the mugearitic character of the rock.

Misleading rock designations could, of course, be avoided by using purely chemical systematics based on the analysis and by using a twofold nomenclature, adding to the rock name, deduced by optical means, an adjective which characterizes its chemical composition.

Attempts to establish a general rock nomenclature in this way (P. NIGGLI, 1931) have not been successful. Suggestions that rocks be defined by numerical symbols (e.g. C.I.P.W. system) have received even less acceptance mainly because the actual meaning of those symbols remain incomprehensible to an ordinary earth-scientist who is not a specialist in calculating and using such symbols.

Equally impracticable are the many rock names which have a local significance, characterizing only a particular rock type found in a specific area. A few examples of such local names are compiled in Table 3.

To Table 3 the comment must be added that several of the magma types defined by P. NIGGLI (1936) actually refer to plutonic rocks which have no molten counterparts, such as the magmatic accumulates, and to some plutonics of metamorphic origin. Such rocks should not be included in the nomenclature of magma types.

3. Systematics of Volcanic Rocks

The proposed calculation from the chemical analysis yields a mineral composition which would have resulted if the crystallization of the rock had been completed either under volcanic or under plutonic conditions.

Such calculations reflect the mineral assemblage stable at the crystallization temperature of the last fraction of the melt and should be based
on the real composition of the rock-forming minerals and not on ideal
normative compounds. Expressing the results in volume percent, the

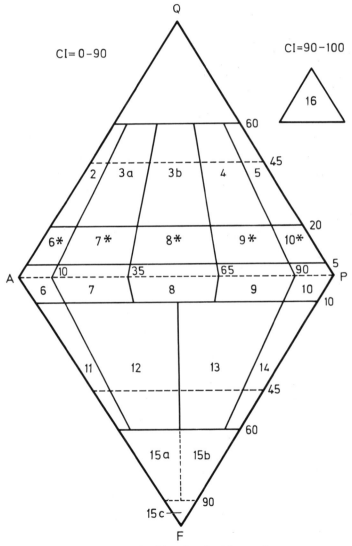

Fig. 3. Q-A-P-F (STRECKEISEN) double-triangle. The rock names corresponding
to each field are compiled in Tables 19 and 20 (pp. 132–139)

plotted point of a volcanic rock in any quantitative scheme of classification determines its correct petrographic name.

The scheme of classification of igneous rocks adopted in this book is based on the double-triangle proposed by P. Niggli (1931). For the classification of volcanic rocks it has been slightly changed (A. Rittmann, 1933, 1960) and finally elaborated by A. Streckeisen (1967, 1968). In its latest form (Fig. 3; Tables 19 and 20 on pp. 132–139) it has been accepted by a large number of petrologists, and its general acceptance is highly desirable in order to put into use, internationally, an uniform nomenclature of volcanic rocks[2].

However, in some of the families, represented by fields in the double-triangle, subdivisions are indicated beyond those based on the colour index. This is particularly needed for the latiandesitic/latibasaltic and andesitic/basaltic rocks (Field 9 and 10). It has become a general practice to distinguish between *tholeiites, alkali basalts* and *high-alumina basalts*. In the first approximation one can state that tholeiites and high-alumina basalts fall in the upper part of Field 10, whereas the undersaturated alkali basalts lie below the line A-P in the lower part of it. A distinction between tholeiites and high-alumina basalts is not possible on the basis of the double-triangle and other criteria are needed.

Recently it has been shown (A. Rittmann, 1967, 1970; V. Gottini, 1968, 1970) that the lavas of all active volcanoes can be divided into two well separated groups which reflect the tectonic situation of the volcanoes. This "bimodality of volcanism" appears very clearly in a diagram the coordinates of which are the value τ (V. Gottini, 1968) and the value σ (serial index of A. Rittmann, 1957) being:

$$\tau = (Al_2O_3 - Na_2O)/TiO_2 \qquad \text{(weight \%)}$$
$$\sigma = (K_2O + Na_2O)^2/(SiO_2 - 43) \quad \text{(weight \%)}$$

as shown in Figs. 4 and 5.

The value τ is suitable to distinguish high-alumina basalts from tholeiites and alkali basalts. In fact, Kuno's high-alumina basalts (Kuno, 1960) have τ values greater than 10 to about 30, whereas the τ values of typical tholeiites (G. A. MacDonald and T. Katsura, 1964) range be-

---

2 T. F. W. Barth (in Streckeisen, 1968) writes of this problem: "The International Union of Geological Sciences (IUGS), realizing the importance of complete and efficient exploitation of the ever increasing body of geological data, has established a number of committees and working groups whose underlying motif is to systematize and bring order into the present confusion... The IUGS realizes that in order to make use of the existing data, we must provide a bridge from the classical forms of geology and their nomenclature, so that quantitative data can be interpreted in geological terms. It is of great importance, therefore, to have as a basis an adequate rock classification and rock nomenclature."

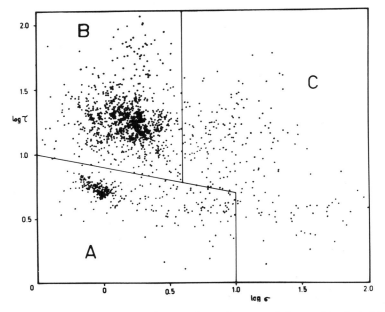

Fig. 4. 1300 analyses of lavas from active volcanoes of the world, plotted in the Gottini-Rittmann diagram. In field *A* are falling only lavas of volcanoes situated in non-orogenic regions, in field *B* those of volcanoes in orogenic belts and island arcs. In field *C* enter the alkaline derivatives of both, i.e. trachytes, phonolites, tephrites etc., among which the sodic types are generally linked to *A*, the potassic ones to *B*

tween 4 and 8, similar to those of alkali basalts. Quite analogous is the distinction between true andesites ($\tau > 10$) and "alkali andesites" to which the name hawaiite has been given (G. A. MacDonald, 1960). Many so-called high-alumina basalts are in reality andesites in the strict sense. Between the high-alumina basalts and the andesites exist all transitions due to a continuous variation of the colour index. It would therefore be better to call the high-alumina basalts *mela andesites*, indicating thus their natural relationship (A. Rittmann, 1971). Some of the so-called tholeiites of Japan are distinguished from the true tholeiites (Hawaii) and tholeiitic flood basalts (cratonic regions) by their high τ values and should be joined with the mela andesites. From the petrological and magmatological point of view the distinction of basaltic and andesitic rocks by means of the τ value is certainly very important. It permits also a clear distinction between true andesites (orogenic) and hawaiites (cratonic alkaline andesites). Using this terminology, andesites *appear* to be limited to orogenic belts, from which the term basalt would

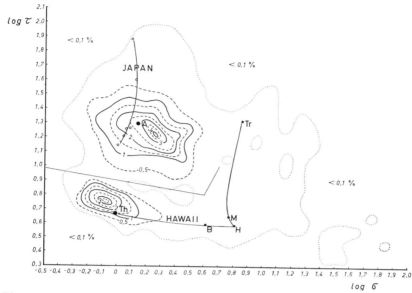

Fig. 5. Position of the high-alumina basalts in GOTTINI's diagram (after A. RITT-
MANN, 1970, p. 980). *A* average of KUNO's high-alumina basalts; Circles = averages of the lavas from Japanese volcanoes (after TANEDA, 1964); *Th* average of
Hawaiian tholeiitic basalts; *B* alkali basalts; *H* hawaiites; *M* mugearites; *Tr* soda-
trachytes. (After G. A. MACDONALD and T. KATSURA, 1961)

be eliminated and confined to cratonic regions. Furthermore, the unfor-
tunate name tholeiite could be replaced by the old name basalt which
would thus be used for the most common types of basaltic rocks
(D. JUNG, 1968). The need for this substitution is all the more indicated,
as for example the original tholeiite of Tholey (Country of Saar,
W-Germany) is not a basaltic rock, but a "hypabyssic, leucocratic rock
of the mangerite family" (D. JUNG, 1958, p. 180) which in the system of
STRECKEISEN is "located in the monzodiorite (latite-andesite) field"
(D. JUNG, 1968, p. 268). Also the term alkali basalt can be criticized
because the prefix "alkali" signifies normally that the feldspar is alkaline
(alkali rhyolite, alkali trachyte etc.), whereas the "alkali basalts" contain
only plagioclase or at least nine times more plagioclase than alkaline
feldspar.

4. Calculation of Stable Mineral Assemblages

In order to calculate accurately the stable mineral assemblage of a
given facies, the average composition of each mineral phase should be
known. In the case of minerals which belong to isomorphic series, the

average composition of their mixed crystals depends upon the chemical composition of the magma from which they separate under the prevailing physical conditions.

The available data which allow us to establish the exact relation between the chemical composition of the magma and that of the mineral constituents of the corresponding volcanic rock are scarce and still insufficient, although a great number of analyses of rocks and of some of their constituents have been carried out.

Notwithstanding the considerable amount of data, it is actually impossible to get all but a very approximate idea about these relationships for the following reasons:

In a number of cases, the analyses of the minerals and those of the bulk of the rock have not been made on the same specimens and, sometimes, even on specimens collected from different lava flows or domes. In other instances the rocks have not been analysed and are only characterized by their names, which for the most part are only phenonames. But even in the relatively few cases where accurate analyses have been carried out on the minerals and on the bulk of the same specimens of rocks, the results may be somewhat misleading. In fact, the mineral analyses have been carried out on phenocrysts, the composition of which hardly ever corresponds to the average composition of that mineral constituent in the rock, which includes also the occasional microlites of the same species. Furthermore, the phenocrysts are for the most part zoned and in separating them by heavy liquids or by a magnetic separator the outermost zones are generally eliminated because they are contamined by other minerals. The material, prepared to be analysed, consists therefore essentially of the cores of phenocrysts chemically different from their mantle and the microlites.

It is to be hoped that, by the use of X-ray and microprobe analysis, better results will be obtained in the near future. However, even with these methods, it will be very difficult to determine the average composition of a given mineral phase in a volcanic rock, even if it is holocrystalline.

Another difficulty arises from the fact that a number of analyses have been made of clinopyroxene phenocrysts in rocks which also contain phenocrysts of orthopyroxene or of hornblende. It is evident that this clinopyroxene does not represent the average clinopyroxene crystallizing under pure volcanic conditions.

In consideration of these facts, the only way to approach the problem is to plot the data in diagrams and find approximate relations between the chemical composition of the rock and the nature of the corresponding clinopyroxenes. The results of this empirical proceeding are the basis of the calculation of clinopyroxenes and amphiboles. It is evident that the real composition of the minerals cannot be obtained accurately in this way, but the approximation is sufficient to distinguish between augite, subcalcic augite, pigeonite, titanaugite, aegirine-augite and aegirine, as well as between hornblende, hastingsite, kaersutite, catophorite, arfvedsonite and riebeckite.

This is sufficient for the use of these calculations for the determination of the rock in any one of the facies. The stable mineral assemblage of the volcanic facies and its interpretation by the means of the double-triangle furnishes also the name of the magma type, whereas the comparison of the stable mineral assemblages of the other facies will give valuable information about heteromorphism.

5. Advantages of Knowing the Stable Mineral Assemblage

The calculation of the stable mineral association of the volcanic facies may serve the following purposes:

1. To fix accurately the position of a particular igneous rock in a quantitative classification scheme of volcanic rocks and to define the correct name of the magma type to which the rock belongs.

2. To characterize a glassy rock or a rock containing a vitreous or microscopically indeterminable groundmass.

3. To disclose possible disequilibria by comparing the calculated stable mineral association with the mode. Such disequilibria may provide information dealing with the process of consolidation and degassing of the magma and, thus, also to reflect indirectly events during its ascent and eruption.

4. In case the volcanic rock contains phenocrysts suspected to be of intratelluric origin, the result of the calculation reveals that the rock belongs to a mixed facies and gives a logical explanation of the difference between it and the mode.

5. Taking the possible rock heteromorphism into account, the result of the calculation permits a mutual control between the chemical analysis and the mineralogical composition of the rock.

6. The calculation of the stable mineral assemblage provides data for estimating the oxygen fugacity in the magma and for estimating the possible secondary oxidation of the magma itself or of the volcanic rock (p. 24—29).

7. Calculating the stable mineral assemblage of the volcanic facies of a subvolcanic or a plutonic rock permits a study of rock heteromorphism (pp. 70—73).

II. Igneous Rock Facies

1. Stability of Igneous Minerals

Every mineral is stable within a certain range of temperature, bulk pressure, chemical composition and water pressure of the magma from which it crystallizes.

The stability fields of stoichiometrically ideal components of silicate systems are well known as far as "dry" melts are concerned. More recent

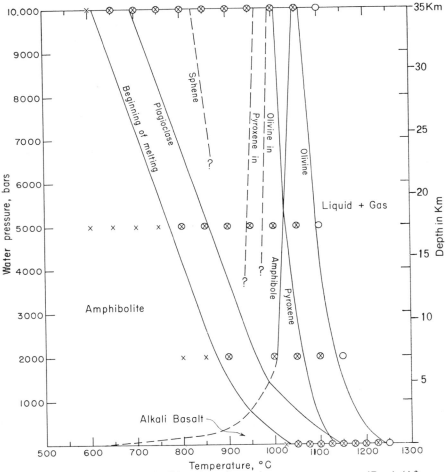

Fig. 6. Crystallization of alkali basalt under various vapour pressure (P_{H_2O}). (After H. S. YODER and C. E. TILLEY, 1962, p. 452)

work has been done on silicate melts under high pressure in presence of water, i.e. under conditions similar to those of natural magmas. The results of these experimental researches are of great importance to the understanding of crystallization processes in natural magmas. However, most of the experiments have been carried out on closed systems with only a few components under idealized conditions.

Natural magmas have a more complicated composition. Hence, the results obtained in the laboratory can be applied to natural magmatic processes only with precaution and after a careful examination of the natural conditions. The presence of magmatic gases lowers the melting points and changes, sometimes in a drastic manner, the range of stability of single mineral phases. Without taking this and similar facts into account, erroneous conclusions may be drawn. Experiments carried out under high water pressure in melts of volcanic rocks have thrown new light on the problems of crystallization of natural magmas. An example, represented in Fig. 6, illustrates the influence of water pressure in an alkali basalt magma (YODER and TILLEY, 1962). At low pressure, olivine, pyroxene and plagioclase will form successively, whereas at high pressure, the most important femic mineral will be an amphibole.

Hydroxyl-bearing minerals, such as amphiboles and micas, can only form if the water pressure exceeds a certain value specific for the mineral in question. At lower water pressures hydroxyl-bearing minerals become unstable and will convert to other minerals which are stable under the prevailing conditions. The following stoichiometrically ideal relations may serve as examples:

High water pressure

$Ca_2Mg_4Al_2Si_7O_{22}(OH)_2 - H_2O$

hornblende − water

$K(Mg, Fe)_3AlSi_3O_{10}(OH)_2 - H_2O$

biotite − water

Low water pressure

$\rightarrow \quad CaAl_2Si_2O_8 + CaMg_4Si_5O_{15}$

$\rightarrow \quad$ anorthite + subcalcic augite

$\rightarrow \quad KAlSi_2O_6 + 2Mg_2SiO_4 + FeO$

$\rightarrow \quad$ leucite + olivine + iron ore

or if sufficient silica is present:

$K(Mg, Fe)_3AlSi_3O_{10}(OH)_2 + 4 SiO_2 - H_2O \quad \rightarrow \quad KAlSi_3O_8 + 5 (Mg, Fe)SiO_3$

biotite + quartz − water $\qquad\qquad \rightarrow \quad$ sanidine + pigeonite

$\qquad\qquad\qquad\qquad\qquad\qquad\qquad$ (in volcanic rocks).

If a magma containing intratelluric phenocrysts of biotite or hornblende is poured out on the earth's surface, most of its gases escape and the phenocrysts become unstable and should be converted into stable minerals as illustrated in the above equations. However, the rapid cooling and consolidation of the melt hinders or even prevents the slow reactions and the phenocrysts are frozen in as a *metastable phase* which is not in equilibrium with the other mineral phases. Strictly speaking, any kind of zoned crystals are also metastable phases due to a high rate of cooling. If the cooling had been slow, the already-formed crystals belonging to an isomorphous series would have had the time to react with the remaining melt to continuously maintain equilibrium among the phases. In this case completely homogeneous crystals would have formed.

2. The Concept of Mineral Facies

The concept of mineral facies has proved to be very useful in the petrography of metamorphic rocks and should be extended to igneous rocks. Theoretically, the term mineral facies comprises the stable mineral assemblage crystallizing from any given kind of magma under definite physical conditions. According to this rigid definition, every time a new mineral becomes stable for a slight variation of pressure or temperature, another mineral facies should be distinguished. In doing so, a rather great number of mineral facies would result. Practically, a more elastic definition must be introduced, saying that an igneous facies is characterized by the assemblage of stable minerals crystallizing from magmas within a certain range of physical conditions.

For our purposes it seems convenient to distinguish the following mineral facies:

1. *Volcanic facies,* including the stable mineral assemblages formed at relatively low pressure and high temperature. According to the water pressure one can distinguish two sub-facies:

a) a "dry" volcanic facies,

b) a "wet" volcanic facies.

2. *Plutonic facies,* including the stable mineral assemblages formed at high pressures. Here also two sub-facies can be distinguished:

a) a "dry" plutonic facies, formed at relatively high temperatures under low water pressure,

b) a "wet" plutonic facies, formed at lower temperatures under high water pressure.

These two facies grade into each other and are linked by stable mineral assemblages formed under subvolcanic conditions. Most of these sub-volcanic assemblages are very similar to those of the "wet" volcanic facies with which they will be treated.

If, under subvolcanic conditions, the partial pressure of carbon dioxide is high, carbonate minerals will crystallize directly from the magma. The resulting stable mineral assemblage characterizes a particular facies, called the *carbonatite facies,* which will be treated separately.

3. The Volcanic and Subvolcanic Facies

A volcanic rock is considered to represent the *pure volcanic facies* if the crystallization of the lava, largely degassed at low pressure, proceeded slowly enough to maintain chemical equilibrium among the single phases to the very end of the crystallization process.

Such an ideal consolidation of the lava is rare. The cooling is nearly always too rapid to permit completion of the reactions between the crystallized constituents and the remaining melt. The chemical adjustments of the equilibrium required by the decreasing temperature are generally slower than the rate of cooling. Therefore, single crystals of minerals representing solid solutions of variable bulk composition (plagioclase, clinopyroxene, olivine, etc.) are mostly zoned. Minerals of reaction pairs (olivine/pyroxene) occur in amounts not corresponding to an equilibrium at the end point of crystallization. In addition, in many cases, the time available for crystallization is insufficient, causing a partial consolidation of the lava as metastable glass.

The *pure subvolcanic facies* comprises mineral assemblages stable at pressures of the magmatic gases (mainly H_2O) high enough to affect notably the temperature range of crystallization. The gaps in solid solutions between certain minerals will be more extended. The stability field of the high temperature phases of reaction pairs will be reduced, sometimes to such a degree that the corresponding low temperature phase can melt congruently, i.e. it will crystallize directly from the melt which is in equilibrium with it. The relatively high vapour pressure causes the crystallization of hydrated femic minerals like micas. This facies is best represented by dike rocks which crystallize at moderate depth. There are, however, also lava flows with mineral assemblages corresponding to the subvolcanic facies. Thus, for instance, many trachytes and rhyolites contain microlites of biotite which have surely crystallized after the eruption of the molten mass at the earth's surface. Such lavas have apparently consolidated under subvolcanic conditions. How can this pseudo-subvolcanic behaviour be explained?

It is a well known fact to volcanologists that even under volcanic conditions, i.e. under very low bulk pressure, the escape of gases from the lava may be strongly hindered if the viscosity of the melt is high. In very viscous lavas the gas pressure may reach very high values. This is demonstrated by the violent outburst of glowing clouds from lava domes. The explanation of the pseudo-subvolcanic conditions must be looked for in the degree of viscosity. At a given temperature viscosity is much greater in acid magmas than in basic ones. With decreasing temperature the viscosity increases exponentially. Table 4 gives some examples of this behaviour.

Table 4. Temperature and viscosity of some lavas (viscosity in poises)

Basalts	$t = 1000°$ C	viscosity $= 10^4$ to 10^5
Andesites	$t = 1000°$ C	viscosity $= 10^6$ to 10^7
Dacites	$t = 1000°$ C	viscosity $= 10^9$

Mela andesites from Oo-Shima (after T. MINAKAMI and S. SAKUMA, 1953, p. 93)

1951, March	$t = 1125°$ C	viscosity $= 5.6 \times 10^3$
1951, March	$t = 1083°$ C	viscosity $= 7.1 \times 10^4$
1951, March	$t = 1038°$ C	viscosity $= 2.3 \times 10^5$

Basalts from Mauna Loa and Kilauea, Hawaii (after G. A. MACDONALD, 1963)

1952	$t = 1100$ to $1050°$ C	viscosity $= 3 \times 10^4$
1950	$t = 1070$	viscosity $= 4 \times 10^3$

On the other hand, acid and salic magmas crystallize at much lower temperatures than basic ones. This fact leads to the conclusion that basic magmas will produce mineral assemblages corresponding closely to the pure "dry" volcanic facies, whereas acid and salic ones will crystallize under pseudo-subvolcanic conditions. Thus, the volcanic facies presents characteristic features of both the pure volcanic and the pure subvolcanic facies which are grading into each other.

4. The Plutonic Facies

The stable mineral assemblages of the "dry" plutonic facies are formed under high bulk pressure, relatively high temperature and very low water pressure. In many respects stable minerals of this facies are similar to those of the "dry" volcanic facies insofar as no hydrated silicates can be formed. At high pressure, very light minerals in particular are not stable if they can be replaced by heavier ones. For instance, leucite having the density 2.47 will be unstable at high pressure.

The "wet" plutonic facies is characterized by high bulk pressure and high water pressure which lowers the temperature of crystallization. Such conditions are realized at a depth where the cooling of an intruded magma is extremely slow, so that the crystallizing phases will always be in equilibrium with the rest melt. The resulting holocrystalline coarse-grained rocks will be constituted of homogeneous crystalline phases, i.e. without any zoning. Owing to the high water pressure hydrated silicates like micas and amphiboles will be stable.

The plutonic facies treated in this book concerns only rocks of magmatic origin. However, the petrogenetic convergence of igneous plutonites and metamorphic plutonites of similar chemical composition permits the calculation of the stable mineral assemblages of both types by the aid of the methods proposed in this book. There are, however, many rocks, said to be plutonic rocks, the chemical composition of which does not correspond to any known volcanic rock. Among these the following types may be mentioned: dunites, peridotites, pyroxenites, biotitites, anorthosites as well as italites, etc., which are crystal cumulates due to gravitative differentiation or kata-metamorphites. Even contact-metamorphic products, like skarns, have been classified as plutonites. Such rocks of non-magmatic origin cannot be calculated by the proposed methods (see Appendix p. 183, "Keys for Ultramafic Rocks ...").

5. The Carbonatite Facies

Carbonatites form under subvolcanic conditions if the partial pressure of carbon dioxide is sufficiently high to cause the crystallization of calcite and other carbonate minerals directly from the magma. In basic magmas, rich in lime and magnesia and poor in silica, the amount of calcite or dolomite may become very large. The carbonatite facies grades into pneumatolitic and hydrothermal mineral deposits. In this book, only magmatic carbonatites will be treated.

Table 5. Stable minerals in the various facies (Vd = "dry" volcanic facies, Vw = "wet" volcanic facies, Pd = "dry" plutonic facies, Pw = "wet" plutonic facies, C = carbonatitic facies)

Mineral	Vd	Vw	Pd	Pw	C
Quartz	+	+	+	+	0
Sanidine	+	+	0	0	0
Anorthoclase	+	+	0	0	0
Orthoclase	0	0	+	+	+
Plagioclase	+	+	+	+	+
Leucite	+	+	0	0	0
Nepheline	+	+	+	+	+
Kalsilite	+	+			0
Sodalite group	+	+	0	+	+
Orthopyroxenes	0	+	+	+	
Augite	+	+	+	+	+
Subcalcic augite	+	0	0	0	0
Pigeonite	+	+	0	0	0
Aegirine-augite	+	+	+	+	+
Amphiboles	0	0	0	+	0
Biotite	0	+	0	+	+
Muscovite	0	+	0	+	0
Olivine	+	+	+	+	+
Melilite	+	+	0	0	0
Melanite	0	+	0	+	0

The common accessories: apatite, magnetite, ilmenite, spinels, etc. are ubiquitous.

6. Stable Minerals of the Various Facies

In Table 5 the stable minerals are indicated by the sign +, the unstable ones by the sign 0. If these latter appear modally in a given facies they must be metastable. The facies are indicated by the following symbols: Vd = "dry" volcanic facies, Vw = "wet" volcanic facies, Pd and Pw = "dry" and "wet" plutonic facies, C = carbonatite facies.

7. Mixed Facies

The mixed facies of a volcanic rock is characterized by containing intratelluric phenocrysts of the plutonic or subvolcanic facies and a groundmass of the volcanic facies. Phenocrysts of biotite, amphiboles, hypersthene or, in some alkaline rock types, melanite are metastable under

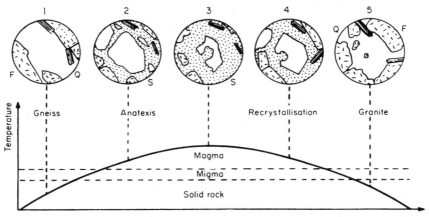

Fig. 7. Anatectic melting and recrystallization of quartz and feldspars. In the lower part of the diagram, the variation in temperature in the buried sial is indicated, the highest temperature being reached at about the middle of the field. In the upper part of the diagram, the states of the granitic materials corresponding to each temperature are portrayed. *1* Starting material: gneiss with granoblastic quartz (*Q*), feldspar (*F*) and biotite. *2* Anatectic magma with partly melted crystalloblasts. The melt (*S*) has roughly alkali rhyolitic composition. *3* Anatectic magma with largely melted crystalloblasts. The quartz appears strongly "corroded". *4* The anatectic magma cools. Quartz and feldspar are partly recrystallized, but the deepest embayments in the quartz are not yet healed over. *5* End product: orthogranite with hypidiomorphic granular texture. On complete crystallization, the quartz is unable to retain its crystal form and again makes up anhedral interstitial masses. If the anatectic magma is erupted, the melt solidifies to glass or to an extremely fine-grained crystalline groundmass in which, according to the moment at which the eruption occurred, quartz and feldspar phenocrysts of stages *2*, *3* or *4* are embedded.

(After A. RITTMANN, 1962, p. 202)

volcanic conditions and hence not in equilibrium with the groundmass. Their real instability is often demonstrated by their more or less transformation into aggregates of minerals stable in the volcanic facies.

Variations of the prevailing physical conditions in subvolcanic and plutonic igneous rocks can also lead to the formation of mixed facies. If a magma of intermediate composition with a moderate water pressure cools down slowly in depth, the early formed crystals will be anhydrous in accordance with the conditions of the "dry" plutonic facies. The originally small amount of magmatic gases will remain in the rest melt, the water pressure of which may increase to such a degree that it will cause the crystallization of hydrated silicates, typical of the "wet" plutonic facies. The resulting rock will display a mixed facies.

Volcanic rocks produced by anatectic magmas may contain partially fused crystals of minerals originating from metamorphic rocks. Many so-called phenocrysts of quartz in rhyolites, rhyodacites or dacites, exhibiting shapes characteristic of corroded crystals, are actually relict xenocrysts from rocks subjected to partial anatexis (Fig. 7; A. RITTMANN, 1958). Xenocrysts of orthoclase (low temperature optics) and the rarer ones of plagioclase can easily be taken as true magmatic phenocrysts, especially in cases where the relictic xenocrysts form the core surrounded by shells of the same minerals (high temperature optics) crystallized later from the anatectic magma. In such cases, however, the crystals often exhibit anomalous kinds of zoning, with abrupt changes in composition, and corroded cores. Most xenocrysts, and particularly those of garnet, cordierite and other minerals formed in depth, are not in equilibrium with the groundmass formed under volcanic conditions. In these cases also the rock presents a mixed facies.

III. Basic Principles of the Calculation

In contrast to the C.I.P.W. norm and to the Kata-norm of NIGGLI, both based on simple stoichiometric compounds, the present calculation is essentially empirical in character and uses average compositions of the actual minerals. These average compositions of the constituent minerals are, of course, not constant throughout all rocks but vary with the rock type. This variation, in other words the relationship between the chemical composition of the magma (bulk analysis of the rock) and those of the constituent minerals crystallized out of it, must be known in adequate approximation, assuming equilibrium conditions. For this purpose, analytical data of the rock-forming minerals and for host rocks were statistically evaluated. The composition of a constituent, say clinopyroxene, found to be characteristic of a certain rock type, was used in the calculation of the stable mineral assemblage of that particular rock.

1. Unit of Calculation

To facilitate the calculation, a proposal by P. NIGGLI (1936) must be adopted. According to him not the chemical formula of a mineral but the single electropositive element contained in it is used as a unit. The quantity of a single compound or of a more complex chemical unit is expressed through the number of positive ions present, neglecting the negative ions (O, S, Cl, F). However, in contrast to NIGGLI's proposal and in order to facilitate certain calculations, also complex negative ions such as CO_3^{2-} and SO_4^{2-} will be counted. The following examples illustrate this manner of expressing quantities:

$$\text{K-feldspar} \quad KAlSi_3O_8 = \ 5 \ Or$$
$$3 \ \text{K-feldspar} \quad 3KAlSi_3O_8 = 15 \ Or \ .$$

The simple reaction between leucite and quartz to form K-feldspar expressed stoichiometrically:

$$KAlSi_2O_6 + SiO_2 = KAlSi_3O_8$$

will be written

$$4 \ Lc + 1 \ Q = 5 \ Or \ .$$

Taking the average composition of natural leucite into account this reaction equation must be modified to:

$$K_{10}NaAl_{11}Si_{22}O_{66} + 11\ SiO_2 = K_{10}NaAl_{11}Si_{33}O_{88}$$
$$44\ Lc + 11\ Q\quad = 55\ Sanidine\ (San).$$

A more complex unit like a clinopyroxene has the following composition:

$$(Na_{0.06}\ Ca_{0.84}\ Mg_{0.10})\ (Mg_{0.40}\ Fe_{0.34}\ Ti_{0.06}\ Al_{0.20})\ (Al_{0.26}\ Si_{1.74})\ O_{6.00}.$$

This formula can be expressed in the pyroxene components wollastonite, enstatite, ferrosilite, Tschermak's component and jadeite as follows:

$$64\ CaSiO_3 + 50\ MgSiO_3 + 28\ FeSiO_3 + 10\ CaAl_2SiO_6 + 6\ NaAlSi_2O_6$$

which, divided by 2, can be written in symbols:

$$32\ Wo + 25\ En + 14\ Fs + 20\ Ts + 3\ Jd.$$

On the other hand, one unit of clinopyroxene will always be:

$$1\ Cpx = 1/4\ (X\ Y\ Z_2)$$

no matter what ions occupy the positions of X, Y and Z. Likewise, the unit of a feldspar (sanidine, anorthoclase or plagioclase) is 1/5 of the stoichiometric formula; that of leucite is 1/4, that of nepheline 1/3, that of olivine 1/3, that of melilite 1/5, that of calcite 1/2 of the respective formulae.

2. Number of Atoms

The first step in the calculation consists in transforming the weight percentages of oxides given in the analysis into numbers of atoms. For this purpose, the weight percentage of an oxide which contains only one positive ion is divided by the respective molecular weight, whereas the weight percentages of an oxide containing two positive ions must be divided by one half of the molecular weight (Table 6).

To avoid decimals, it is convenient to multiply the results by 1000, rounding off to integer numbers. In step 1 of Key 1 (p. 88), reciprocal values of the molecular or half molecular weights are given to facilitate the calculation.

The negative ions, substituting oxygen, must also be transformed into numbers of atoms, but, like oxygen, will not be included in the figure indicating the quantity of the compound:

$$S : 32.06 = S;\quad Cl : 35.46 = Cl;\quad F : 19.00 = F.$$

H_2O+ and H_2O- are neglected.

Table 6. Molecular weights per number of electropositive atoms

SiO_2	: 60.06 = Si	TiO_2	: 79.90 = Ti
Al_2O_3	: 50.97 = Al	P_2O_5	: 70.98 = P
Fe_2O_3	: 79.84 = Fe^{3+}	BaO	: 153.36 = Ba
FeO	: 71.84 = Fe^{2+}	SrO	: 103.63 = Sr
MnO	: 70.93 = Mn	NiO	: 74.69 = Ni
MgO	: 40.39 = Mg	Cr_2O_3	: 76.01 = Cr
CaO	: 56.08 = Ca	ZrO_2	: 123.22 = Zr
Na_2O	: 30.99 = Na	SO_3	: 80.06 = SO_3
K_2O	: 47.10 = K	CO_2	: 44.01 = CO_2

3. Degree of Oxidation

The degree of oxidation of a rock, denoted Ox^0, is defined as

$$Ox^0 = Fe^{3+}/(Fe^{3+} + Fe^{2+} + Mn)$$

and is calculated from the numbers of these atoms.

The degree of oxidation is always lower in the magma than in the solidified volcanic rock. This circumstance arises from the fact that, in the course of the crystallization, water will become enriched in the rest melt and will advance the oxidation. Therefore, in a volcanic rock, the femic phenocrysts (including magnetite if present), representing an early stage of crystallization, will reflect a lower oxidation degree than the last microlitic crystallizations or the bulk of the groundmass.

In rocks belonging to a magmatic differentiation series, the average oxidation degree increases towards the more salic differentiates. As a consequence, a positive correlation could be anticipated between the degree of oxidation and the alkali content of rocks. In fact, such a statistical correlation has been found to exist. For example, it is well illustrated by the series of alkali basalts, hawaiites and mugearites from Hawaii, as is shown in Fig. 8 by the line which may be expressed numerically by the equation:

$$Ox^0 = 1.2 \, [0.1 + (Na + K) \cdot 10^{-3}] \, .$$

In magmas of the calc-alkaline series, i.e. in andesites, dacites and rhyolites, the average degree of oxidation is greater, as is shown in Fig. 9 by the straight line. For such rocks, e.g. of Japan, the following equation is valid:

$$Ox^0 = 1.8 \, [0.07 + (Na + K) \cdot 10^{-3}] \, .$$

In these rocks also the oxidation degree is greater than in the corresponding magmas. However, it may be taken for granted that the oxygen

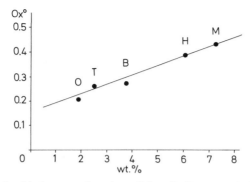

Fig. 8. Degree of oxidation as a function of the alkali content in Hawaiian lavas. Averages: *O* oceanites, *T* tholeiitic basalts, *B* alkali basalts, *H* hawaiites, *M* mugearites

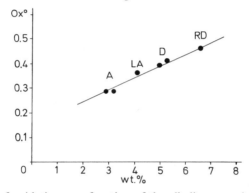

Fig. 9. Degree of oxidation as a function of the alkali content in Japanese lavas. Averages: *A* mela andesites and andesites, *LA* latiandesites, *D* dacites, *RD* rhyodacites

fugacity is generally greater in the magmas of orogenic belts than it is in magmas of cratonic regions. This average oxidation degree of the rocks is certainly greater than that prevailing in the magma from which they solidified. The lowest oxidation degree observed in some specimens of these rocks is about one half of the average value and probably reflects the oxidation degree of magmas before their solidification sets in.

4. Formation of Magnetite

The degree of oxidation controls the formation of magnetite during the consolidation of the lava. Two generations of magnetite can often be distinguished in volcanic rocks:

1. Separate micro-phenocrysts or inclusions in phenocrysts of augite, olivine, etc.

2. Opaque powdery crystallization in the groundmass or in the glass phase causing a dark brown colour or even a submetallic luster of the lava.

The second magnetite generation evidently results from a successive oxidation of the melt either because some water has accumulated in higher levels of the pyromagma or because the top part of the magma column stayed in contact with the air in a molten lava lake or during a long-lasting persistent volcanic activity.

An exceedingly strong oxidation may result from autopneumatolysis characteristic of the topmost part of a consolidating lava mass. A later oxidation of the solidified rock is caused by secondary alteration. A quite noticeable oxidation may also take place during grinding of the specimen for chemical analysis.

Taking all the factors into account which affect the oxidation degree of a rock, the following conclusions may be drawn:

1. The quantitative distinction between the ferrous and the ferric iron in a chemical analysis is not sufficient to warrant a correct estimation of

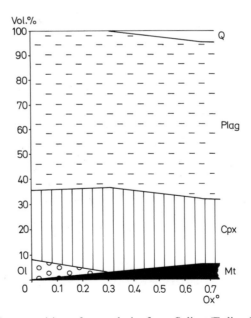

Fig. 10. Mineral composition of an andesite from Salina (Eolian Islands/Italy) as a function of oxygene fugacity

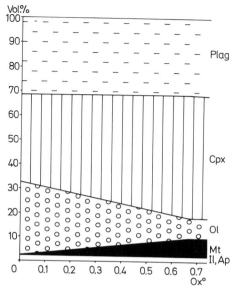

Fig. 11. Mineral composition of an oceanite from Hawaii as a function of oxygene
fugacity

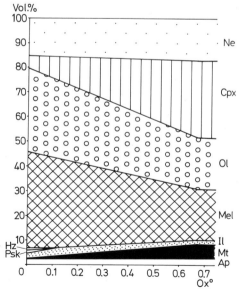

Fig. 12. Mineral composition of a melilitite from Homboll (Hegau/W.-Germany)
as a function of oxygene fugacity

the oxygen fugacity in the magma, even though the specimen was ground under special precautions.

2. If the analysis indicates a high degree of oxidation, it must be determined whether the specimen was really free from secondary alteration. The calculation of the analysis of an altered rock is of no magmatological significance. However, for readers who wish to check a partially altered rock, a scheme of calculation is given in the appendix (p. 182).

3. Because the iron content in ferromagnesian silicates is known to be partially in the ferric state, the calculation of magnetite (Mt) or hematite (Hm) merely on the basis of the figure for Fe_2O_3 indicated by the analysis is not justified.

4. For determining the magma type to which the rock belongs and for the purpose of comparing the stable mineral assemblage of two similar rocks with each other, the same value for Ox^0 must be adopted for both rocks. In other words, the calculation of magnetite must be standardized. Figs. 10, 11, and 12 illustrate the necessity of standardizing the degree of oxidation. The calculated mineral assemblages of the rocks selected depend strongly on the value for Ox^0 used in the calculation.

5. Saturated Norm

The saturated norm is obtained by distributing the numbers of atoms among saturated silicates, accessory minerals and occasional compounds, without regard to the available amount of Si. In doing so, there remains a difference between the available Si and the Si' which has been used to form the saturated compounds:

$$\Delta Q = Si - Si' .$$

If ΔQ is positive in sign, the rock is oversaturated and in its saturated norm quartz will appear. If ΔQ is negative it is reported as such to indicate a silica deficiency.

The saturated norm serves as a basis for the calculation of the stable mineral assemblages for all facies of igneous rocks.

For calculating the saturated norm, the following rules must be applied in the given order:

a) Calculation of Occasional Components

If CO_2, SO_3, Cl, F, S or ZrO_2 are given in the analysis, the following minerals or compounds are calculated using the amounts of electronegative elements or of the anhydrides as units.

CO_2: distinguish two alternatives:
  a) $CO_2 < Ca$.
  Calcite $Cc = 2\ CO_2$ according to the formula $CaCO_3$.
  Remainder: $Ca^* = Ca - CO_2$.

b) $CO_2 \geqq Ca$.

Calcite $Cc = 2\ Ca$. Remainder: $CO_2{}^* = CO_2 - Ca$.

Breunnerite $Br = 2\ CO_2{}^*$ according to the formula $(Mg,Fe)CO_3$.
Remainder: $Ca = 0$; $(Mg,Fe)^* = (Mg,Fe) + Ca - CO_2$.

SO_3: distinguish two alternatives:

    a) $SO_3 < Ca$.

Anhydrite $Ah = 2\ SO_3$, according to the formula $CaSO_4$.
Remainder: $Ca^* = Ca - SO_3$.

    b) $SO_3 \geqq Ca$.

Anhydrite $Ah = 2\ Ca$. Remainder: $SO_3{}^* = SO_3 - Ca$.

Thenardite $Th = 3\ SO_3{}^*$, according to the formula Na_2SO_4 .
Remainder: $Ca^* = 0$; $Na^* = Na - 2\ SO_3{}^*$.

Cl: *Halite* $Hl = Cl$, according to the formula NaCl.
Remainder: $Na^* = Na - Cl$.

F: No special compound is calculated. For convenience the excess of fluorine over that contained in apatite, biotite or amphibole is not considered.

Zr: *Zircon* $Z = 2\ Zr$, according to the formula $ZrSiO_4$.
Remainder: $Si^* = Si - Zr$.

S: *Pyrite* $Pr = 0.5\ S$, according to the formula FeS_2.
Remainder: $Fe^* = Fe - 0.5S$.

b) Calculation of Common Accessories

In all types of saturated norms, apatite and ilmenite will always be calculated in the same manner:

Apatite $Ap = 2.667\ P$, according to the formula $Ca_5P_3O_{12}(F,Cl)$.
Remainder: $Ca^* = Ca - 1.667\ P$.

Ilmenite $Il = 2\ Ti$, according to the formula $FeTiO_3$.
Remainder: $Fe^* = Fe - Ti$. [3]

Magnetite. The formation of magnetite is controlled by the degree of oxidation. Theoretically, if $Fe^{3+} < 2\ Fe^{2+}$ and if $(Na + K) < Al$, then all Fe^{3+} should enter magnetite, and magnetite should equal $1.5 \times Ox^0 \times$ total iron.

Actually some Fe^{3+} enters femic silicate minerals, as illustrated in Figs. 13, 14 and 15. On the basis of statistical evaluation of appropriate data, the average degree of oxidation which prevails in the magma can be assumed to amount to almost two thirds of the value valid for the corresponding rock. That means that the early-formed silicates are in

---

[3] It may happen that $Ti > Fe^{2+}$. If so, then add to Fe^{2+} an amount of Fe^{3+} equal to $(Ti - Fe^{2+})$ in order to form Il. This adjustment is justified by the fact that all volcanic rocks appear more oxidized than the magmas from which they originated.

equilibrium with an amount of magnetite corresponding to the Ox^0 of the magma. Hence, the number of Fe atoms entering magnetite will be obtained by multiplying the average Ox^0 of the rocks by the number of Fe-atoms available after the formation of ilmenite:

$$Mt_0 = \overline{Ox^0}\,(Fe - Ti)\,.$$

On p. 24 two equations have been given for Ox^0, one for basaltic magmas and their derivatives, the other for andesitic to rhyolitic magmas. In order to establish which one of the two must be applied, the value of Il (ilmenite) must be used as the discriminating factor. In fact, the basic and intermediate Hawaiian magmas, and similar ones all over the world, are less oxidized and contain much more TiO_2 than the more oxidized and TiO_2-poor Japanese and analogous lavas. On the basis of these considerations, the following equations will be used for rocks in which $(Na + K) < Al < (2\,Ca + Na + K)$:

If $Il > 25$, then $Mt_0 = 1.2\,[0.1 + (Na + K) \cdot 10^{-3}]\,(Fe - Ti)$,
if $Il \leqq 25$, then $Mt_0 = 1.8\,[0.07 + (Na + K) \cdot 10^{-3}]\,(Fe - Ti)$.

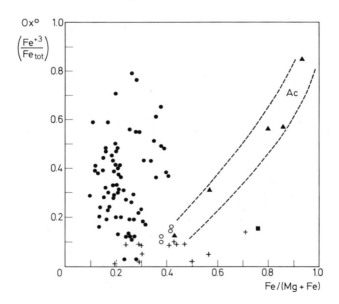

Fig. 13. Degree of oxidation in clinopyroxenes. Abscissa: $Fe/(Mg + Fe)$ $(Fe = $ total iron atoms), ordinate: $Fe^{3+}/Fe = Fe^{3+}/(Fe^{2+} + Fe^{3+})$. Signs: crosses = pigeonites, circles = sodian augites, triangles = aegirine and aegirine-augites, dots = other clinopyroxenes (augite, titanaugite, etc.)

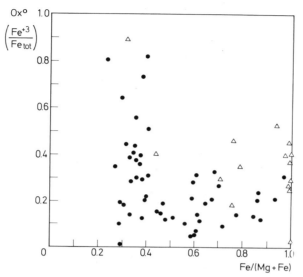

Fig. 14. Degree of oxidation in amphiboles. Abscissa: Fe/(Mg + Fe) (Fe = total iron atoms), ordinate: $Fe^{3+}/Fe = Fe^{3+}/(Fe^{2+} + Fe^{3+})$. Signs: circles = calcic amphiboles, triangles = alkali amphiboles

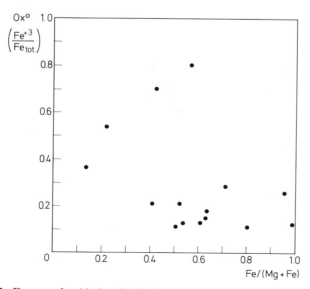

Fig. 15. Degree of oxidation in biotites. Abscissa: Fe/(Mg + Fe) (Fe = total iron atoms), ordinate: $Fe^{3+}/Fe = Fe^{3+}/(Fe^{2+} + Fe^{3+})$

For peraluminous rocks with $Al > (2\,Ca + Na + K)$ the equation reads:

$$Mt_0 = [0.10 + (Na + K) \cdot 10^{-3}](Fe - Ti)$$

and for alkaline rocks with $Al \lessgtr (Na + K)$:

$$Mt_0 = [0.1 + (Na + K) \cdot 10^{-3}]\,(Fe + Na + K - Al - Ti)\,.$$

In peralkaline rocks with $(Na + K) > (Al + Fe^{3+})$ no Mt_0 will form. According to the relative amounts of $(Na + K)$ and Al, the amount of available Fe multiplied by the factor a will yield the standard magnetite Mt_0. There are three alternatives:

a) $(Na + K) \leqq Al$: $\qquad\qquad Mt_0 = a \cdot (Fe - Ti)$
b) $Al < (Na + K) \leqq (Al + Fe^{3+})$: $\ Mt_0 = a \cdot (Fe + Al - Ti - Na - K)$
c) $(Na + K) > (Al + Fe^{3+})$: $\qquad Mt_0 = 0\,.$

c) Calculation of Saturated Silicates

After having formed the accessories and occasional compounds, the remaining amounts of atoms are distributed among the following normative silicate minerals (Table 7):

Table 7. Saturated silicate molecules

Quartz	SiO_2	$= 1\ Q$
Orthoclase	$KAlSi_3O_8$	$= 5\ Or$
Albite	$NaAlSi_3O_8$	$= 5\ Ab$
Anorthite	$CaAl_2Si_2O_8$	$= 5\ An$
Sillimanite	Al_2SiO_5	$= 3\ Sil$
Wollastonite	$CaSiO_3$	$= 2\ Wo$
Enstatite	$MgSiO_3$	$= 2\ En$
Ferrosilite	$FeSiO_3$	$= 2\ Fs$
Acmite	$NaFeSi_2O_6$	$= 4\ Ac$
K-Acmite	$KFeSi_2O_6$	$= 4\ K\text{-}Ac$
Sodasilite	Na_2SiO_3	$= 3\ Ns$

Some of the silicates exclude each other, namely *Sil* and *Wo*, *An* and *Ac* (or *K-Ac*), *(Sil + An)* and *Ns*, and also Mt_0 and *Ns*. Consequently according to the various amounts of atoms, five types of saturated norms can be distinguished (Table 8):

Table 8. The five types of saturated norms

1	2	3	4	5
Ap	Ap	Ap	Ap	Ap
Il	Il	Il	Il	Il
Mt_0	Mt_0	Mt_0	—	—
Or	Or	Or	Or	Or
Ab	Ab	Ab	Ab	—
An	An	—	—	—
Sil	—	—	—	—
—	Wo	Wo	Wo	Wo
En	En	En	En	En
Fs	Fs	Fs	Fs	Fs
—	—	Ac	Ac	Ac
—	—	—	—	K-Ac
—	—	—	Ns	Ns
ΔQ	ΔQ	ΔQ	ΔQ	ΔQ

These five types of saturated norms are conditioned by the following relations among the available amounts of atoms:

1st type: $Al > (2 Ca + Na + K - 3.33 P)$,

2nd type: $(Na + K) < Al \leqq (2 Ca + Na + K - 3.33 P)$,

3rd type: $Al < (Na + K) \leqq (Al + Fe^{3+})$,

4th type: $(Na + K) > (Al + Fe^{3+})$ and $K < Al$,

5th type: $K > Al$.

The second type is by far the most frequent one. The third type occurs in soda rhyolites, soda trachytes and phonolites. Very sodic types of these rocks may yield norms of the fourth type, whereas the fifth type of norm is extremely rare. The first type results from peraluminous rocks such as cordierite granite or cordierite rhyolite, etc.

Stress must be laid on the fact that altered rocks which have lost alkalis also may yield the first type of saturated norm. Secondary argillaceous products and sericite will appear as Sil in the norm.

Detailed indications for calculating the saturated norms are given in Key 1 (pp. 88–91).

Besides their fundamental importance for calculating stable mineral assemblages, saturated norms are often useful for illustrating petrochemical characteristics of rock series. The following suggestion may be made:

Calculate the ratio $T = An/Il$ and plot it as ordinate against the abscissa ΔQ, as illustrated in Fig. 16. The value of T presents a discriminative capacity similar to that of τ (p. 9) to distinguish tholeiites and alkali basalts from the so-called high-alumina basalts or hawaiites from true andesites.

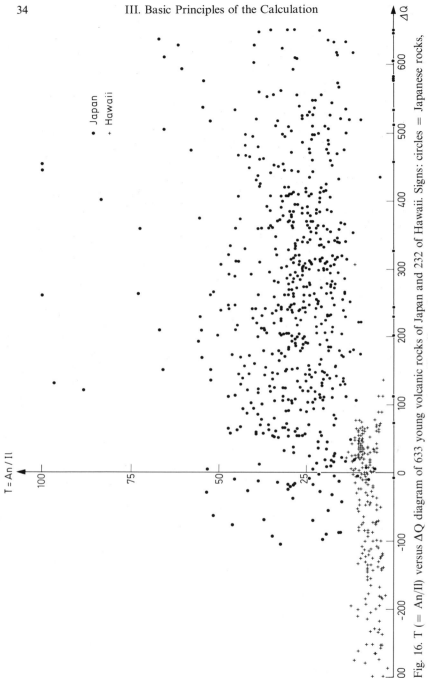

Fig. 16. T (= An/Il) versus ΔQ diagram of 633 young volcanic rocks of Japan and 232 of Hawaii. Signs: circles = Japanese rocks, crosses = Hawaiian rocks

IV. Igneous Rock Forming Minerals

1. Introductory Remarks

Most rock forming minerals represent mixed crystals of variable composition which is a function of the composition of the magma from which they originate and of the prevailing physical conditions during crystallization.

On the basis of available data, the determination of these relations is possible only in an approximation which is sufficiently accurate to permit the calculation of the quantity of a given mineral, but not of its qualitative composition. It is easy to calculate the amount of hypersthene, but not the ratio Mg/Fe in it. For that reason, often in the following calculations only the compound *Hy* will be used instead of its components *En* and *Fs*. This substitution will by no means prevent the determination of the amounts of stable minerals or the correct denomination of the rock and of the corresponding magma type.

For convenience, the chemical composition of the minerals will be expressed in components of the saturated norm. In doing so, it is possible to subtract the components of a given mineral, member by member, from the saturated norm of the rock, as illustrated in the following example of a nepheline basanite:

Sat. norm	Ap	Il	Mt_0	Or	Ab	An	Wo	Hy	ΔQ	sum
Rock	21	88	37	95	520	397	192	732	−315 =	1767
−Augite	−	−26	−	−1	−17	−44	−192	−192	+22 =	−450
Remainder:	21	62	37	94	503	353	−	540	−293 =	1317

The remaining amounts of Ap, Il and Mt are final. The other components will be distributed among olivine, nepheline and feldspars as shall be shown later in detail (pp. 61–66). The amount of augite (450) in the above example is determined by the fact that all available *Wo* will enter into it.

2. Chemically Simple Minerals and Accessories

These minerals can be calculated according to their stoichiometrical formulae for two reasons: either they are very pure and their composition actually corresponds to their formulae so closely that even if they occur in considerable amounts no significant error will be committed, or they are present in so small amounts that the error, introduced by neglecting their impurities, will be negligible as far as it concerns the determination of their quantity.

Some of these minerals are already represented among the compounds of the saturated norm (Table 9): Among these, only the amounts of apatite, calcite, zircon and pyrite given in the saturated norm will be contained as such in the final result of the calculation. Most others enter, at least partially, into minerals of more complex composition.

Table 9. Simple normative accessory minerals

Apatite	$Ca_5P_3O_{12}(F,Cl)$	$= 8\ Ap$
Ilmenite	$FeTiO_3$	$= 2\ Il$
Magnetite	Fe_3O_4	$= 3\ Mt_0$
Calcite	$CaCO_3$	$= 2\ Cc$
Anhydrite	$CaSO_4$	$= 2\ Ah$
Thenardite	Na_2SO_4	$= 2\ Th$
Halite	$NaCl$	$= 1\ Hl$
Zircon	$ZrSiO_4$	$= 2\ Z$
Pyrite	FeS_2	$= 1\ Pr$
Sodasilite	Na_2SiO_3	$= 3\ Ns$

The occasional remainders, however, will be quoted as such.

Quartz will appear only if, after the formation of all silicates, some silica is left over. Then $Q = \Delta Q$.

Other simple minerals do not appear in the saturated norm. Their composition will be represented by adding or subtracting several members of the saturated norm (Table 10).

Table 10. Chemically complex minerals expressed in combined norms

Spinel	$MgAl_2O_4$	$= 3\ Sp = 2\ En + 3\ Sil - 2\ Q$
Pleonast	$(Mg,Fe)Al_2O_4$	$= 3\ Sp = 2\ Hy + 3\ Sil - 2\ Q$
Hercynite	$FeAl_2O_4$	$= 3\ Hc = 2\ Fs + 3\ Sil - 2\ Q$
Ulvöspinel	Fe_2TiO_4	$= 3\ Usp = 1\ Mt + 2\ Il$
Perovskite	$CaTiO_3$	$= 2\ Psk = 2\ Wo + 2\ Il - 2\ Fs$
Sphene (Titanite)	$CaTiSiO_5$	$= 3\ Tn = 2\ Wo + 2\ Il + 1\ Q - 2\ Fs$
Cossyrite	$Na_4Fe_9Ti_2Si_{11}O_{37}$	$= 26\ Coss = 6\ Ns + 4\ Il + 14\ Fs + 2\ Q$
Cordierite	$(Mg,Fe)_2Al_4Si_5O_{18}$	$= 11\ Cd = 6\ Sil + 4\ Hy + 1\ Q$
Garnet	$(Fe,Mn)_3Al_2Si_3O_{12}$	$= 8\ Gr = 6\ Fs + 3\ Sil - 1\ Q$

Some remarks on opaque minerals should be added: In some rocks, *magnetite* has an almost stoichiometric composition. In other rocks it may contain considerable quantities of Ti. Such *titanomagnetites* are transitional types between magnetite and ulvöspinel. Their composition can be expressed as

$$m \cdot Mt_0 + n \cdot Il, \quad being \quad m > 0.5n.$$

In ultrabasic rocks (e.g. melilitites) some Fe^{3+} may be replaced by Al, corresponding to mixtures of Mt_0 and Hc.

In the following calculations, magnetite, ilmenite and hercynite will be given separately. They may be added, if the actual opaque mineral is known to be titanomagnetite, ulvöspinel or Al-bearing magnetite. *Chromite* will not appear as such but as magnetite, because Cr has been added to Fe^{3+}. However, a glance at the numbers of atoms will immediately show if sufficient Cr is available to justify the denomination chromite instead of magnetite.

3. Feldspars

The feldspars contain three components: Or, Ab and An. In terms of these components the average composition of feldspars may be expressed by the formula

$$Or_x Ab_y An_z \text{ in which } x + y + z = 100\%.$$

The three components are not miscibile in all proportions but exhibit a gap in solid solubility the width of which varies with the temperature as illustrated in Fig. 17. The temperature of crystallization is affected by the water pressure in the magmas: the higher the gas content the lower is the crystallization temperature. The four triangular diagrams differ from each other mainly in the stability field of anorthoclase[4] which diminishes with decreasing temperature and disappears at about 720° C. Simultaneously, the composition of the alkali feldspars and of the plagioclases which coexist with each other are shifted to some extent as shown by lines of coexistence in Fig. 17. As a consequence, the quantitative ratio between alkali feldspar and plagioclase present in the rock will also be shifted with the temperature.

The continuous variation in the amounts and compositions of the feldspars cannot be taken rigorously into account in the calculation. On the basis of statistical studies, dealing with the feldspars in igneous rocks (BARTH, 1956), two diagrams are presented which illustrate the coexistence and composition of the feldspars in the igneous rock facies.

---

4 The term anorthoclase is used to indicate the chemical composition of the feldspar and not as a mineral name (see W. E. TRÖGER, 1969, pp. 678/679).

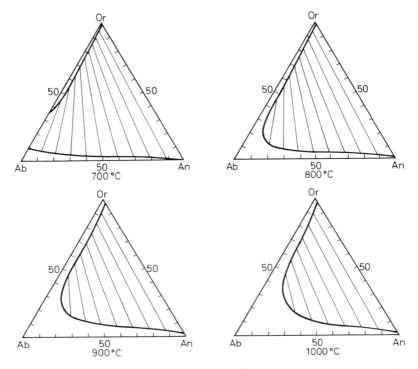

Fig. 17. Four feldspar triangles for various temperatures. (According to data
published by T. F. W. Barth, 1956, p. 363)

For the more potassium-rich sanidine varieties and for the more basic
plagioclases the diagram for the pure volcanic facies (see Fig. 49)
corresponds to a temperature of 900° C, and for the more acid feldspars
to a temperature of 800° C. Under the conditions of the subvolcanic
(resp. pseudo-subvolcanic) facies these temperatures are about 50° lower,
and in that of the "wet" plutonic facies (see Fig. 62) they correspond to
about 800° and 700° C. respectively. Of course, all intermediate cases
also exist, but referring only to these two average diagrams the results of
calculating the stable feldspars are immediately comparable with each
other.

The amounts of feldspar components in the saturated norm of the
rock do not represent the actual average feldspar because part of them
enters potentially into the clinopyroxenes, micas and amphiboles; and,
in undersaturated rocks, alkaline feldspars may be converted, partly or
completely, into feldspathoids. Only the amounts of the components

Or*, Ab* and An* which are left over after the formation of the mentioned minerals will form feldspars, the average composition of which will be $Or_x^* Ab_y^* An_z^*$.

4. Nepheline and Kalsilite

Nepheline is stable in all facies of igneous rocks. Its chemical composition never corresponds exactly to the theoretical formula $NaAlSiO_4$. About one third of sodium can be replaced by potassium. According to the available data the alkali ratio $k = K/(Na + K)$ in nepheline is about one third of that of the bulk analysis. Its maximum value in volcanic rocks may reach 0.30, but in "wet" plutonic rocks only 0.15. Hence the crystallization of nepheline causes an increase of the alkali ratio in the remaining melt, leading (in strongly undersaturated magmas) to a potassium enrichment high enough to cause the formation of leucite after that of nepheline. In such cases, nepheline forms euhedral phenocrysts, whereas leucite will appear as anhedral fillings of the interstices.

In addition, especially in the less undersaturated igneous rocks, the nepheline structure contains silica in excess over the stoichiometric ratio. In volcanic rocks containing only small amounts of nepheline the excess of silica may be as much as 8 mol-%, whereas in some nephelinites and in nepheline melilitites the mineral corresponds to the formula $(Na, K)AlSiO_4$. In potassium-rich nephelines the silica excess is much smaller and becomes nil, if the alkali ratio k is greater than 0.4.

Kalsilite does not contain excess of silica, but averages 9 mol-% $NaAlSiO_4$. Hence the formula of average kalsilite corresponds to $K_{10}NaAl_{11}Si_{11}O_{44}$.

5. Leucite

The average composition of leucite corresponds to the formula $K_{10}NaAl_{11}Si_{22}O_{66}$; i.e. it contains 9 mol-% $NaAlSi_2O_6$. The range of variation in the actual ratio of potassium to sodium in leucite is slight and can be neglected for the purpose of calculating the stable mineral assemblage.

Leucite occurs only in volcanic and subvolcanic rocks but never in true plutonites because its stability field shrinks rapidly with increasing pressure, as illustrated schematically in Fig. 18.

Euhedral crystals of leucite occur only in relatively potassium-rich volcanic rocks ($k \geq 0.4$), but anhedral leucite may form also from more strongly undersaturated sodic magmas after enrichment of potassium in the melt in consequence of the crystallization of nepheline (see above).

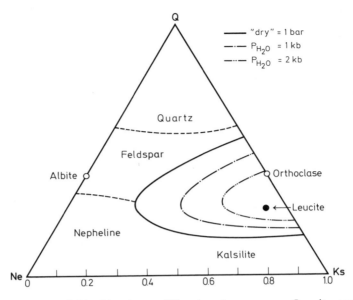

Fig. 18. Stability field of leucite at different water pressures. Leucite occurs in undersaturated lavas only if $k > 0.40$ and in viscous lavas only if $k > 0.58$
$$[k = \mathrm{Or}/(\mathrm{Or} + \mathrm{Ab})]$$

The opposite may occur in strongly undersaturated potassic magmas in which the separation of leucite causes an enrichment of sodium in the remaining melt to such a degree that anhedral nepheline will form after the crystallization of leucite. Some authors have believed that from undersaturated magmas with alkali ratios greater than 0.4 leucite will form in any case. Statistics, however, show that the problem is not as simple as that. In fact, there are plenty of undersaturated lavas with k-values between 0.4 and 0.58 which do not contain any modal leucite. It was therefore necessary to search for another discriminating factor than the alkali ratio. On the basis of the following reasons such a factor has been found.

During volcanic eruptions most magmatic gases escape into the air and only small amounts are retained in the outflowing lavas. It is a well known fact that femic lavas (basalts, tephrites, etc.) are very fluid and therefore lose their gases easily, whereas salic lavas (trachytes, phonolites, etc.) retain their gases owing to their high viscosity. The trapped gases must reach high tensions to overcome the internal friction of the melt and to escape in an explosive manner. It is evident that the gases in viscous salic lavas are under relatively high pressure, thus causing

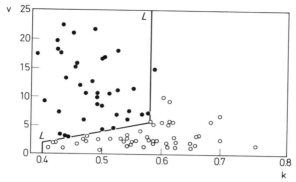

Fig. 19. Stability of leucite and nepheline as a function of $k - v'$. Abscissa: $k =$ Or/(Or + Ab), ordinate: $v' = $ (Or + Ab)/(Wo + Hy). Signs: dots = nepheline, circles = leucite

conditions comparable to those of the subvolcanic facies. Under these conditions the stability field of leucite must be considerably reduced. What particular feldspathoid will crystallize from a magma depends not only on the alkali ratio and on the degree of undersaturation, but also on the viscosity of the melt.

The viscosity depends, in its turn, on the chemical composition of the melt. It is low in femic melts but high in salic ones, not only because of their chemical composition but also because they are much cooler. In a way, the ratio $v = $ (Or + Ab)/(Wo + Hy) must be roughly proportional to the viscosity, and herewith is the discriminating factor looked for.

Plotting the values v of this ratio against the alkali ratio k, a graph should result in which the fields of formation of leucite and of nepheline should appear separated. Fig. 19 shows that this is really the case. For the values $k = 0.4$ to 0.58 there appears a clear boundary (L) between the field of rocks resulting from the crystallization of basic fluid magmas with phenocrysts of leucite, and that of rocks derived from viscous magmas without phenocrysts of leucite. The boundary line between the two fields can be defined by the equation $L = 20k - 6$.

Concluding, it can be stated that:

if k is smaller than 0.2, only nepheline will form,

if k lies between 0.2 and 0.4, nepheline will crystallize first and only in highly undersaturated rocks anhedral leucite may form later.

If k lies between 0.4 and 0.58, either nepheline or leucite will crystallize first, according to the degree of viscosity, expressed by the ratio v. If $v < L$, then leucite will form first, if $v > L$, then nepheline will crystallize first.

If $k = 0.58$ to 0.91, leucite will form first and, if the magma is highly undersaturated, anhedral nepheline may be formed afterwards.

If $k > 0.91$, only leucite can be formed.

On the basis of these statements, the fields of stability of nepheline and leucite can be represented as a function of the alkali ratio and the degree of silica deficiency (q^*). Fig. 20 illustrates these stability fields for fluid magmas $v < L$ (cf. Fig. 19), and Fig. 21 for viscous magmas $v > L$ (cf. Fig. 19).

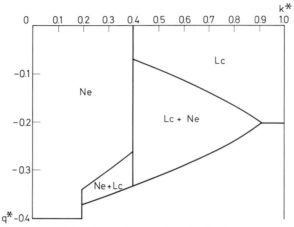

Fig. 20. Formation of leucite and nepheline in fluid magmas as a function of k^* and q^*. Abscissa: $k^* = Or^*/(Or^* + Ab^*)$, ordinate: $q^* = \Delta Q^*/(Or^* + Ab^*)$

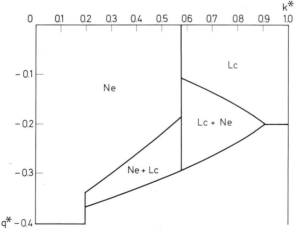

Fig. 21. Formation of nepheline and leucite in viscous magmas as a function of k^* and q^*. Abscissa: $k^* = Or^*/(Or^* + Ab^*)$, ordinate: $q^* = \Delta Q^*/(Or^* + Ab^*)$

6. Sodalite Group

Sodalite, nosean and haüyne form solid solutions with each other, usually with one of these components predominating. Sodalite is often nearly ideal in composition. In haüyne part of the sodium may be replaced by potassium as in nepheline. Because these three feldspathoids are present only in relatively small amounts, the simple stoichiometrical formulae may be used for their calculation:

Sodalite $6\,NaAlSiO_4 \cdot 2NaCl$ $= 9\,Ne + 1\,Hl$,

Nosean $6\,NaAlSiO_4 \cdot Na_2SO_4$ $= 9\,Ne + 1\,Th$,

Haüyne $6\,(Na,K)AlSiO_4 \cdot CaSO_4 = 9\,Ne + 1\,Ah$.

These minerals occur only in undersaturated rocks containing Cl and/or SO_3.

7. Cancrinite

Occurs only in undersaturated subvolcanic to plutonic rocks (foyaites) and is mostly of deuteric origin. Its composition corresponds nearly to the formula

$$6\,NaAlSiO_4 \cdot CaCO_3 = 9\,Ne + 1\,Cc .$$

8. Scapolites

In saturated and oversaturated rocks, instead of sodalite minerals, scapolites may be found.

They are essentially mixed crystals of

Marialite $6\,NaAlSi_3O_8 \cdot 2NaCl = 15\,Ab + 1\,Hl$,

Mejonite $6\,CaAl_2Si_2O_8 \cdot CaCO_3 = 15\,An + 1\,Cc$.

Marialite is very rare in lavas and will not be considered in the calculation. The other members exist only in subvolcanic and plutonic rocks.

9. Pyroxenes

In the pure volcanic facies only clinopyroxenes (Cpx) are stable. Occasional phenocrysts of orthopyroxenes (Opx) have been formed under subvolcanic or plutonic conditions and are not in equilibrium with the phases of the groundmass. In more acid and salic lavas the gases remain partly trapped in the viscous melt which crystallizes at a lower temperature under pseudo-subvolcanic conditions. In this case, and more so under

Table 11. Saturated norms of pyroxenes

Pyroxenes	Il	Or	Ab	An	Ac	Ns	Wo	En	Fs	ΔQ	Sum	Ox°	Deer et al., 2 Page/No.
Salite	7	13	39	107	—	—	764	765	201	−86	1810	0.332	50 / 16
Salite	59	12	110	346	—	—	583	627	194	−150	1781	0.206	51 / 23
Aegirine	19	20	161	—	1530	5	—	—	24	−22	1732	0.974	82 / 3
Aegirine	16	15	268	—	1275	61	117	29	92	−73	1744	0.855	83 / 6
Aegirine-augite	64	19	103	—	1073	—	184	197	—	64	1765	0.906	83 / 8
Aegirine-augite	40	—	126	—	835	—	341	213	178	2	1735	0.753	83 / 9
Aegirine-augite	10	41	174	—	549	—	478	263	292	−81	1726	0.608	84 / 10
Aegirine-augite	19	23	268	—	13	—	742	441	374	−131	1749	0.404	84 / 12
Augite	12	2	44	128	—	—	634	794	228	−43	1799	0.144	115 / 6
Augite	31	7	108	193	—	—	671	720	244	−173	1801	0.325	115 / 7
Augite	14	—	37	96	—	—	636	720	296	−24	1775	0.130	116 / 11
Augite	33	8	58	87	—	—	607	707	322	−49	1773	0.094	117 / 16
Augite	15	3	39	87	—	—	550	625	464	−37	1746	0.082	118 / 22
Sodian augite	29	4	448	36	—	—	669	520	252	−173	1785	0.377	119 / 28
Ferroaugite	30	29	168	72	—	—	586	361	586	−137	1695	0.036	120 / 7
Ferroaugite	18	4	37	31	—	—	661	175	760	−33	1653	0.047	121 / 12
Subcalcic augite	60	13	61	59	—	—	369	649	491	12	1714	0.097	122 / 2
Subcalcic ferroaugite	3	21	94	132	—	—	231	341	778	18	1618	0.149	122 / 5
Titanaugite	73	10	145	308	—	—	675	582	159	−180	1772	0.481	123 / 1
Titanaugite	96	—	58	476	—	—	650	386	258	−202	1722	0.378	123 / 3
Titanaugite	108	19	126	257	—	—	690	556	140	−128	1768	0.401	123 / 4
Titanaugite	138	14	171	268	—	—	674	527	116	−145	1763	0.435	123 / 5
Pigeonite	5	—	31	6	—	—	142	1166	507	−49	1808	0.051	146 / 1
Pigeonite	16	—	18	74	—	—	167	928	610	−50	1763	0.033	146 / 4
Pigeonite	15	24	42	36	—	—	237	799	649	−55	1746	0.004	146 / 5
Pigeonite	13	2	3	50	—	—	134	737	552	19	1710	0.001	147 / 9
Pigeonite	21	13	37	19	—	—	128	629	823	0	1670	0.051	147 / 11
Bronzite	5	—	—	75	—	—	12	1341	443	−26	1850	—	17 / 8
Hypersthene	5	—	44	—	25	—	95	1181	507	−45	1812	0.062	17 / 10
Hypersthene	3	7	48	—	21	—	27	1072	661	−58	1781	0.049	18 / 12
Hypersthene	2	—	—	21	—	—	37	920	760	3	1743	0.023	18 / 15
Ferrohypersthene	4	1	2	17	—	—	18	643	1002	−17	1670	—	19 / 18

plutonic conditions, the miscibility gap grows wider and two pyroxenes, Cpx and Opx, coexist in equilibrium. This behaviour is illustrated in the examples on pp. 142 and 147–148.

The components of *orthopyroxenes* are En and Fs. The amounts of other components, like Wo, are generally so small that they may be neglected.

On the other hand the composition of *clinopyroxenes* is often very complex as exemplified by the saturated norms of Cpx in Table 11.

In addition to the essential components Wo, En, Fs, and occasional Ac, considerable amounts of An, Ab, Or and Il are usually present in Cpx. Theoretically, the normative feldspar components should be desilicated to yield the pyroxene components of Tschermak's molecule (Ts), jadeite (Jd) and "potassian jadeite" (K-Jd) according to the equations:

$$5\,An - 1\,Q = 4\,Ts\,,$$
$$5\,Ab - 1\,Q = 4\,Jd\,,$$
$$5\,Or - 1\,Q = 4\,K\text{-}Jd\,,$$

i.e. the silica deficiency $-\Delta Q$ should equal $-0.2(Or + Ab + An)$. Actually, $\Delta Q = -0.35(Or + Ab + An)$ on the average. Marking the components of the saturated norm entering clinopyroxene with a dash, the general norms of clinopyroxenes will be:

$$Cpx = Il' + Or' + Ab' + An' + Wo' + Hy' - Q'$$

resp. for alkaline pyroxenes:

$$Cpx = Il' + Or' + Ab' + Ac' + Wo' + Hy' - Q'\,.$$

The relative amounts of the single components of a given clino-pyroxene depend, naturally, upon the composition of the magma from which the Cpx crystallizes. On the basis of the available data, the following approximate relations have been established:

1. The amount of feldspar components $(Fsp' = Or' + Ab' + An')$ in Cpx is about proportional to that of Wo for a given ΔQ of the saturated norm of the rock, let F' be the factor of proportionality, i.e. $Fsp' = F' \cdot Wo$. F' can be expressed by the following equations:

If $\Delta Q > O$, then $F' = 0.2\,(1 - \Delta Q \cdot 10^{-3})$.
If $\Delta Q < O$, then $F' = 0.2 - \Delta Q \cdot 10^{-3}$.

2. The relative amounts of the single Fsp' components are roughly proportional to the values Or, Ab and An of the rock. Let x, y and z be the factor of proportionality, then the relations are:

$$Or' : Ab' : An' = x\,Or : y\,Ab : z\,An\,.$$

Here also the relations are somewhat different in oversaturated and undersaturated rocks, namely:

If $\Delta Q > 0$: then $x = 0.3$ Or, $y = 1.0$ Ab, $z = 2.0$ An.
If $\Delta Q < 0$: then $x = 0.2$ Or, $y = 0.6$ Ab, $z = 2.0$ An.

Let be $D = F' \cdot Wo/(x + y + z)$, then the feldspar components in Cpx will be:

$$Or' = x\,D; \quad Ab' = y\,D; \quad An' = z\,D\,.$$

In most instances, these components will be much smaller than the corresponding members of the saturated norm of the rock. However, in Wo-rich undersaturated rocks it may happen that An' > An, and even that Ab' > Ab. In such cases put An' = An resp., Ab' = Ab. The sum Fsp' will then be smaller than F'· Wo. Hence, the determination of Q' must be based in all cases on the relation: $Q' = -0.35\,(Or' + Ab' + An')$.

3. The usually small amounts of Il' in the Cpx norm naturally depends upon the available Il in the rock, but there is no direct proportionality. Several attempts to find a factor "p" of proportionality have shown that it increases with Wo and decreases with An and with the alkali ratio:

$$k = K/(K + Na) = Or/(Or + Ab).$$

The best relation found is the following:

If $k > 0.4$, then $p = 0.3$ Wo/An.
If $k < 0.4$, then $p = 0.6$ Wo/An,
being: $Il' = p \cdot Il$.
The upper limit of p is, naturally, 1, i.e. $Il' = Il$.

4. In the volcanic and the dry plutonic facies, all available Wo and all Ac will enter clinopyroxene. In oversaturated rocks in pure volcanic facies, all of the Hy also enters Cpx, whereas in undersaturated rocks some Hy may be converted into olivine, if $Hy > n \cdot Wo$. If, on the contrary, $Hy < n \cdot Wo$, all Hy enters Cpx and no olivine can be formed.

The numerical value of n has been shown to depend on the value of ΔQ and of k, according to the following approximate relations:

If $\Delta Q \geq 0$ and $k < 0.4$, then $n = 1.3$.
If $\Delta Q \geq 0$ and $k \geq 0.4$, then $n = 1.2$.
If $\Delta Q < 0$ and $k < 0.4$, then $n = 1.3 + \Delta Q \cdot 10^{-3}$.
If $\Delta Q < 0$ and $k \geq 0.4$, then $n = 1.2 + \Delta Q \cdot 10^{-3}$.

In the subvolcanic and in the plutonic facies, the maximum amount of Hy' in Cpx (without Ac) equals $n \cdot Wo$. If $Hy > n \cdot Wo$, then a part of Hy, namely $Hy^* = Hy - n \cdot Wo$, will form either hypersthene or will enter

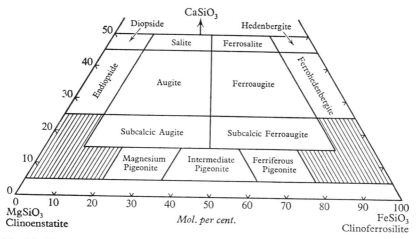

Fig. 22. Nomenclature of clinopyroxenes in the system $CaMgSi_2O_6$–$CaFeSi_2O_6$–$Mg_2Si_2O_6$–$Fe_2Si_2O_6$. (After A. POLDERVAART and H. H. HESS, 1951, p. 472)

biotite and/or amphibole. Sodian augite and aegirine-augite are never associated with olivine or with hypersthene. Hence, in saturated norms of Ac-bearing alkaline Cpx, the ratio Hy/Wo can vary without limits. The pyroxenes can be divided into several types according to their chemical composition (e.g. POLDERVAART and HESS, 1951). On the basis of their saturated norm one may distinguish in igneous rocks the following pyroxenes (Fig. 22).

A. *Cpx with An:* Let be $c = (Wo + 0.4\,An)/Hy$ and split Hy′ into Fs and En (En = 2 Mg):

c	$En > Fs$	$En < Fs$
> 0.9	diopside	hedenbergite
0.5 to 0.9	augite	ferroaugite
0.3 to 0.5	subcalcic augite	subcalcic ferroaugite
0.1 to 0.3	pigeonite	ferropigeonite.

Augites and diopsidic augites ($c' > 0.5$) containing more than 4 % Il in their norm (atoms) are called *titanaugites*.

B. *Cpx with Ac* ($=$ alkaline Cpx): Let be $a = Ac/(Ac + Wo + Hy)$

$\qquad\qquad a \geqq 0.70 \qquad$ aegirine,

$\qquad 0.70 > a \geqq 0.15 \qquad$ aegirine-augite,

$\qquad 0.15 > a \geqq 0 \qquad$ sodian augite.

C. *Opx:* Let be $m = En/(En + Fs)$

$m \geq 0.90$ enstatite,
$0.90 > m \geq 0.70$ bronzite,
$0.70 > m \geq 0.50$ hypersthene,
$0.50 > m \geq 0.30$ ferrohypersthene,
$0.30 > m \geq 0.10$ eulite.

It must be noted that the value of m depends upon the oxidation degree which in turn conditions the amount of Fe entering magnetite and hence, that of Fs. Having standardized the amount of Mt_0, that of Fs has also been standardized so that the type of orthopyroxene will remain unchanged whatever the actual oxidation degree of the analysed rock may be (see also Fig. 10).

10. Olivines

The composition of rock forming olivines corresponds to the formula $(Mg, Fe)_2SiO_4$ so closely, that olivine may be calculated stoichiometrically. Undersaturated rocks of the pure volcanic or of the dry plutonic facies will contain olivine, if in their saturated norm $Hy > n \cdot Wo$, i.e. if, after the calculation of the pyroxene, Hy* has been left over. Opx is converted into olivine according to the equation:

$$4\,Hy - 1\,Q = 3\,Ol\,.$$

The amount of olivine depends upon the relative amounts of the remaining Hy* and the remaining silica deficiency $-\Delta Q^*$.

If $|\Delta Q^*| < 0.25\,Hy^*$, then $Ol = 3|\Delta Q^*|$.
If $|\Delta Q^*| > 0.25\,Hy^*$, then $Ol = 0.75\,Hy^*$.

In the latter case, some silica deficiency $-\Delta Q_1$ will remain:

$$|\Delta Q_1| = |\Delta Q^*| - 0.25\,Hy^*\,.$$

11. Micas

Micas, being hydrated silicates, are stable only under "wet" igneous conditions.

Muscovite of magmatic origin forms only from very few magmas containing a great excess of Al, i.e. in fresh rocks having $Sil > 1.32\,Hy$ in their saturated norm. The average composition of muscovite (Ms) can be expressed with sufficient approximation by the following relative amounts of compounds:

$$100\,Ms = 61\,Or' + 12\,Ab' + 40\,Sil' - 13\,Q'\,.$$

Table 12. Saturated norms of micas

Il	Or	Ab	An	Sil	En	Fs	ΔQ	Deer et al., **3**, Page/No.	
74	839	81	130	79	667	411	−532	58	1
107	695	113	146	55	632	362	−407	58	2
91	866	105	139	61	594	512	−634	58	3
132	866	84	116	111	508	409	−490	58	4
85	973	34	34	178	456	486	−538	59	6
90	882	81	80	75	395	576	−482	59	7
16	918	247	—	29	408	716	−640	59	8
87	839	129	92	134	421	557	−574	59	9
79	834	164	—	60	339	785	−625	59	10
79	876	24	15	151	210	764	−464	60	11
54	957	202	—[a]	—	47	884	−590	60	12
46	932	63	14	43	49	873	−456	60	14

[a] 37 Ac

It is evident that the saturated norms of altered rocks sometimes contain a considerable amount of Sil because their feldspars are more or less converted into argillaceous minerals. If the analyses of such rocks are calculated like those of fresh rocks, a considerable amount of modally non-existing muscovite will result. A correct calculation is illustrated on pp. 154–156.

Biotite hardly ever corresponds to the theoretical formula $K(Mg, Fe)_3AlSi_3O_{10}(OH)_2$. Mostly, some Si is replaced by Al, some K by Na, and some Fe by Ti. In Table 12 some saturated norms of biotites exemplify the variability of their chemical composition. Let us mark the normative compounds of biotite by a prime symbol, in order to distinguish them from the members of the saturated norm of the rock.

The available data have shown that the ratio $k' = Or'/(Or' + Ab')$ depends upon $k = Or/(Or + Ab)$, approximately according to the following relation $k' = 0.30\,k + 0.70$.

Biotite contains mostly Al in excess, i.e. the norm contains Sil, the amount of which is highest in cases where the saturated norm of the rock itself also contains Sil. The average composition of such an Al-rich biotite can be written as follows:

$$100\,Bi = 53\,k'\,Or' + 53\,(1 - k')Ab' + 13\,Sil' + 60\,Hy' + 5\,Il' - 31\,Q'.$$

This equation may be used for calculating biotite of rocks in the saturated norm of which the ratio $h = Hy/(Sil + Hy)$ is less than 0.82. In rocks with $h \geq 0.82$, the amounts of Sil' and Hy' are those of Sil and Hy of the rock (Fig. 23).

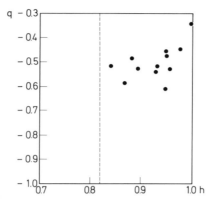

Fig. 23. Biotites in the $h - q$ diagram. Abscissa: $h = \text{Hy}/(\text{Hy} + \text{Sil})$, ordinate: $q = \Delta\text{Q}/(\text{Hy} + \text{Sil})$

Also in rocks which do not contain Sil in their saturated norm, but Wo and An, the biotites show an excess of Al, i.e. some Sil'. For convenience this Sil' will be substituted according to the equation 3 Sil $= 5\,\text{An} - 2\,\text{Wo}$. In such rocks the saturated norm of an average biotite will read as follows:

$$100\,\text{Bi} = 59\,k'\text{Or}' + 59(1\text{--}k')\text{Ab}' + 7\,\text{An}' - 3\,\text{Wo}' + 66\,\text{Hy}' - 29\,\text{Q}'.$$

In rocks containing Ac in the saturated norm, the Al excess in biotites is negligible or actually absent. The average biotite will be given by

$$100\,\text{Bi} = 66\,k'\text{Or}' + 66\,(1 - k')\,\text{Ab}' + 66\,\text{Hy}' - 32\,\text{Q}'.$$

Biotite can be stable in lavas also, if the viscosity is high enough to hinder the escape of gases, whereas in fluid lavas no biotite can form. The viscosity depends upon the chemical composition of the lava and increases exponentially with the decreasing temperature. Acid and salic magmas are cooler and much more viscous than femic ones. Hence, the ratio $v = (\text{Or} + \text{Ab} + \Delta\text{Q})/(\text{Wo} + \text{Hy})$ will be linked in some way to the viscosity, and it may be expected that there exists a limit of the value v above which biotite is stable.

On the other hand, the formation of biotite will depend on the alkali ratio k and on the saturation in silica. Fig. 45 illustrates that the values k and v delimit the stability field of biotite.

The limit can be expressed by the equation: $v = 12 - 12\,k$.

Generally the alkali ratio k must be greater than 0.30 to permit the formation of biotite. However, because the water pressure $P_{\text{H}_2\text{O}}$ and the alkali ratio k increase in the remaining melt during the crystallization of anhydrous silicates, it may happen that towards the end of the crystallization process some biotite may form although the original value of k in the magma was lower than 0.30. Such tardy biotite will not be in equilibrium with the early-formed clinopyroxene. Hence, in calculating the stable mineral assemblage, biotite will form only if k is greater than 0.30.

According to the ratio $m = En/(En + Fs)$ the following varieties of biotite may be distinguished:

Phlogopite $m > 0.80$,
Biotite s.str. $0.80 > m > 0.30$,
Lepidomelane $m < 0.30$.

12. Amphiboles

Even more than clinopyroxenes, amphiboles are extremely variable in composition, as illustrated in Table 13 by the saturated norms of some amphiboles.

On the basis of analyses of rocks and amphiboles the following approximate relations have been deduced:

1. The varieties of amphiboles designated as hornblende, hastingsite, or kaersutite crystallize from magmas the saturated norms of which contain *An* and *Wo*.

2. The composition of these amphiboles varies within certain limits of the ratio $w = Wo/Hy$ as shown in Figs. 24 and 25 and depends, furthermore, upon the degree of silica saturation of the magma and somewhat on the ratio $m = En/Hy$. For convenience of calculation, the most probable limits of the amphibole field are presented by straight lines analytically expressed by the following equations:

In oversaturated rocks: $\Delta Q > 0$
    lower limit: $w_1'' = 0.1 + 0.2\, m$,
    upper limit: $w_2'' = 0.25 + 0.2\, m$.

In undersaturated rocks: $\Delta Q < 0$
    lower limit: $w_1'' = 0.2$,
    upper limit: $w_2'' = 0.4$.

3. The total amount of feldspar compounds (Fsp'') entering these amphiboles depends upon the degree of silica saturation (ΔQ) of the rock, being proportional to the amount of hypersthene compound (Hy'') entering the amphibole. This relation can be expressed by the following

IV. Igneous Rock Forming Minerals

Table 13. Saturated norms of *amphiboles*

Name	Il	Or	Ab	An	Wo	En	Fs	ΔQ	Deer et al., 2, Page/No.	
Hornblende	27	18	306	194	323	787	302	− 808	275	10
Hornblende	8	16	187	363	280	716	311	− 670	276	11
Hornblende	30	75	194	154	344	710	354	− 646	276	15
Hornblende	31	34	106	327	327	660	395	− 616	277	20
Hornblende	32	84	168	92	397	658	479	− 588	278	24
Hornblende	69	152	163	336	270	569	430	− 611	279	26
Hornblende	37	81	156	431	259	516	422	− 470	279	27
Hornblende	39	71	184	103	342	542	497	− 363	279	29
Hornblende	47	71	161	319	263	272	674	− 96	281	36
Hornblende	18	179	81	445	236	138	793	− 5	281	36
Hastingsite	17	13	289	578	133	599	384	− 647	288	13
Hastingsite	39	104	353	455	217	381	523	− 476	289	16
Hastingsite	37	146	292	378	205	220	714	− 188	289	18
Hastingsite	32	207	476	463	190	126	672	− 239	290	22
Hastingsite	16	211	282	336	246	133	801	− 105	290	23
Hastingsite	22	203	332	263	243	66	895	− 23	291	27
Basaltic Hornblende	64	133	366	485	248	716	233	− 1028	317	3
Kaersutite	259	152	616	161	324	640	8	− 990	322	2
Kaersutite	189	63	550	364	270	560	134	− 836	322	4
Kaersutite	175	168	414	402	223	546	147	− 778	322	5
Kaersutite	178	110	544	301	274	557	172	− 824	323	7
Barkevikite	9	67	600	1	418	480	523	− 613	329	1
Barkevikite	6	83	605	198	294	294	706	− 398	329	3

Alkaline amphiboles

Name	Il	Or	Ab	Ac	K-Ac	Ns	Wo	En	Fs	ΔQ	Deer et al., 2, Page/No.	
Riebeckite	24	16	—	732	90	112	7	11	480	+ 105	338	1
Riebeckite	20	69	—	407	89	273	21	15	676	+ 101	338	2
Riebeckite	11	60	—	590	130	18	47	251	442	+ 100	339	6
Arfvedsonite	52	192	255	385	—	178	101	279	427	− 106	367	7
Arfvedsonite	28	106	—	838	41	67	194	65	389	− 41	367	8
Arfvedsonite	12	307	88	497	—	106	73	12	706	− 140	367	9
Catophorite	43	109	269	422	—	76	179	304	440	− 112	360	3
Barkevikite (sodian)	61	101	374	147	—	—	300	282	733	− 314	329	4

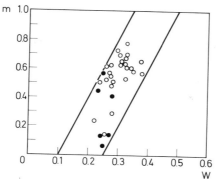

Fig. 24. Amphiboles in the $w-m$ diagram ($\Delta Q > 0$). Abscissa: $w =$ Wo/Hy, ordinate: $m =$ En/Hy

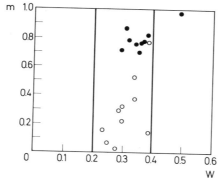

Fig. 25. Amphiboles in the $w-m$ diagram ($\Delta Q < 0$). Abscissa: $w =$ Wo/Hy, ordinate: $m =$ En/Hy

average equations: $f'' \cdot \text{Hy}'' = \text{Fsp}''$ wherein the value of the factor of proportionality f'' is given by:

in oversaturated rocks: $\qquad f'' = 0.5 - 0.4\ \Delta Q \cdot 10^{-3}$,

in undersaturated rocks: $\qquad f'' = 0.5 - 2.33\ \Delta Q \cdot 10^{-3}$.

The ratios among the feldspar compounds are practically the same in the amphiboles and in the rock, i.e.

$$\text{Or}'' : \text{Ab}'' : \text{An}'' = \text{Or} : \text{Ab} : \text{An}.$$

4. In the saturated norm of hornblende or hastingsite the average content of ilmenite is given by the relation

$$\text{Il}'' = 0.03\ \text{Hy}'' \qquad (\text{upper limit Il}'' = \text{Il}).$$

Typical kaersutite occurs in undersaturated rocks and contains more Mg than Fe. In the saturated norm of kaersutite the amount of ilmenite (Il″) increases with $m = En/Hy$ as follows:

$$Il'' = (0.8\,m - 0.37)\,Hy'' \qquad (\text{upper limit } Il'' = Il)\,.$$

On the basis of these relations, the saturated norms of these amphiboles can be calculated:

$$Il'' + Or'' + Ab'' + An'' + Wo'' + Hy'' + Q''\,.$$

According to the ratios of some of these components the amphibole varieties can be defined as follows:

Hornblende: $(Or'' + Ab'' + An'') > 0.75\,Hy''$.
Hastingsite: $(Or'' + Ab'' + An'') < 0.75\,Hy''$ and $Il'' < 0.1\,Hy''$.
Kaersutite: $(Or'' + Ab'' + An'') < 0.75\,Hy''$ and $Il'' > 0.1\,Hy''$.

The alkali amphiboles: *riebeckite, arfvedsonite* and *catophorite* form from magmas whose saturated norms contain *Ac*. Their composition depends upon the degree of silica saturation and on the ratios

$$a'' = Ac''/(Ac'' + Hy'')$$

and

$$w'' = Wo''/(Ac'' + Hy'')$$

as illustrated in Figs. 26 and 27.

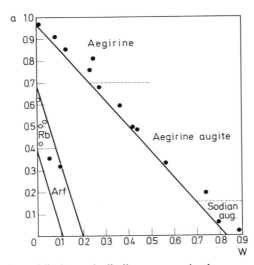

Fig. 26. Alkali amphiboles and alkali pyroxenes in the $w - a$ diagram ($\Delta Q > 0$). Abscissa: $w = Wo/(Ac + Hy)$, ordinate: $a = Ac/(Ac + Hy)$. Rb = riebeckite, Arf = arfvedsonite, Sodian aug. = sodian augite

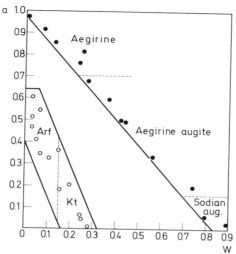

Fig. 27. Alkali amphiboles and alkali pyroxenes in the $w - a$ diagram ($\Delta Q < 0$).
Abscissa: $w = Wo/(Ac + Hy)$, ordinate: $a = Ac/(Ac + Hy)$. Arf = arfvedsonite,
Kt = Catophorite

The limits of the field of alkali amphiboles can be expressed by the following equations:

In oversaturated rocks: $\Delta Q > 0$

$$\text{lower limit} \quad w_1'' = 0.11 - 0.28\, a'',$$
$$\text{upper limit} \quad w_2'' = 0.20 - 0.29\, a''.$$

In undersaturated rocks: $\Delta Q < 0$

$$\text{lower limit} \quad w_1'' = 0.16 - 0.4\, a'',$$
$$\text{upper limit} \quad w_2'' = 0.32 - 0.4\, a'' \qquad \text{being } a'' < 0.55.$$

The following relations can be deduced:

1. The amounts of Ac", Wo" and Hy" available for the formation of alkali amphiboles equal Ac, Wo and Hy of the saturated norm, if amphibole is the only femic silicate. Otherwise they are the remaining parts of Ac, Wo and Hy after the formation of biotite or aegirine-augite or aegirine.

2. The amounts of feldspar components (Or" and Ab") entering alkali amphiboles depend upon the degree of silica saturation (ΔQ) and are proportional to the sum of Ac" + Hy", i.e.:

$$(Or'' + Ab'') = f'' \cdot (Ac'' + Hy'')$$

wherein the factor of proportionality equals:

In oversaturated rocks ($\Delta Q > 0$): $f'' = 0.14 - 2\,a$.

In undersaturated rocks ($\Delta Q < 0$): $\overset{\bullet}{f''} = 0.82 - 1.5\,a$

being $a = Ac/(Ac + Hy)$ of the rock.

Furthermore, $Or'' : Ab'' = Or : Ab$.

Hence $D = (Or'' + Ab'')/(Or + Ab) = f'' \cdot (Ac'' + Hy'')/(Or + Ab)$

and $Or'' = D \cdot Or$ and $Ab'' = D \cdot Ab$.

3. The amount of the silica deficiency (or excess) in alkali amphiboles is found to be

$$Q'' = q'' \cdot (Ac'' + Hy'')$$

wherein in oversaturated rocks ($\Delta Q > 0$):

$$q'' = 1.3\,a\,,$$

in undersaturated rocks ($\Delta Q < 0$):

$$\text{if } a < 0.4 : \qquad q'' = 1.4\,a - 0.51,$$
$$\text{if } a > 0.4 : \qquad q'' = 0.05\,.$$

4. The amount of ilmenite in alkali amphiboles is small:

$$Il'' = 0.02\ Wo'' \qquad (\text{upper limit } Il'' = Il)\,.$$

The various types of alkali amphiboles in igneous rocks can be distinguished as follows:

Riebeckite: $\Delta Q > 0$ and $a > 0.4$.
Arfvedsonite: $\Delta Q > 0$ and $a < 0.4$,
                    $\Delta Q < 0$ and $w < 0.15$.
Catophorite: $\Delta Q < 0$ and $w > 0.15$.

13. Melilites

Melilites can be considered as a mixture of the following components:

åkermanite	$Ca_2(Mg,Fe)Si_2O_7$	$= 5\ \text{Åk}$
gehlenite	$Ca_2Al_2SiO_7$	$= 5\ \text{Geh}$
soda-melilite	$NaCaAlSi_2O_7$	$= 5\ \text{Nm}$
soda-ferrimelilite	$NaCaFeSi_2O_7$	$= 5\ \text{Nfm}$.

The components of melilites appear to be desilicated components of pyroxenes as illustrated by the following relations:

$$2\,CaSiO_3 \; + MgSiO_3 - SiO_2 = Ca_2MgSi_2O_7$$
$$4\,Wo \quad + \;\; 2\,En \;\; - \; 1\,Q \; = 5\,\text{Åk}$$

$$CaAl_2SiO_6 + CaSiO_3 - SiO_2 = Ca_2Al_2SiO_7$$
$$4\,Ts \quad + \;\; 2\,Wo \; - \; 1\,Q \; = 5\,Geh$$

$$NaAlSi_2O_6 + CaSiO_3 - SiO_2 = 5\,NaCaAlSi_2O_7$$
$$4\,Jd \quad + \;\; 2\,Wo \; - \; 1\,Q \; = 5\,Nm$$

$$NaFeSi_2O_6 + CaSiO_3 - SiO_2 = NaCaFeSi_2O_7$$
$$4\,Ac \quad + \;\; 2\,Wo \; - \; 1\,Q \; = 5\,Nfm\,.$$

The desilication of diopside yields olivine in addition to åkermanite:

$$4\,CaMgSi_2O_6 - 3\,SiO_2 = 2\,Ca_2MgSi_2O_7 + Mg_2SiO_4\,,$$
$$16\,Di \quad - \;\; 3\,Q \; = \quad 10\,\text{Åk} \quad + 3\,Fo\,.$$

Some authors consider melilite as belonging to the feldspathoid group. According to the above relations, melilite may rather be called "augitoid". The only statistically significant difference in compo-

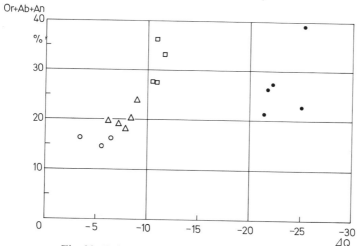

Fig. 28. Relation between clinopyroxenes and melilites
Abscissa: ΔQ $\Big\}$ in % of the saturated norms of these minerals.
Ordinate: Or + Ab + An
Signs : circles = augites in leucitites, triangles = titanaugites in nephelinites, squares = augites in melilite-bearing rocks, dots = melilites in volcanic rocks

Table 14. Saturated norms of melilites

Il	Or	Ab	An	Ac	Wo	En	Fs	ΔQ	Deer et al., 1, Page/No.	
1	45	448	—	117	1245	456	32	-507	242	2
—	183	373	—	67	1147	468	172	-594	242	3
—	21	579	16	—	1233	392	92	-505	242	4
—	35	605	3	—	1185	371	132	-516	242	5
—	83	850	17	—	1058	234	186	-511	243	8

sition between certain clinopyroxenes and melilites in strongly under-saturated rocks is the silica content (Fig. 28, Table 14). The calculation of melilites will be based essentially on the conversion of already calculated clinopyroxene into melilite.

On the other hand, melilites can be considered as clinopyroxenes with additional lime derived from calcite:

$$CaMgSi_2O_6 + CaCO_3 = Ca_2MgSi_2O_7 + CO_2$$
$$4\,Di \quad + \quad 1\,Cc \quad = \quad 5\,Åk \quad + \text{expelled } CO_2$$

$$CaAl_2SiO_6 + CaCO_3 = Ca_2Al_2SiO_7 + CO_2$$
$$4\,Ts \quad + \quad 1\,Cc \quad = \quad 5\,Geh \quad + \text{expelled } CO_2$$

$$NaAlSi_2O_6 + CaCO_3 = NaCaAlSi_2O_7 + CO_2$$
$$4\,Jd \quad + \quad 1\,Cc \quad = \quad 5\,Nm \quad + \text{expelled } CO_2$$

$$NaFeSi_2O_6 + CaCO_3 = NaCaFeSi_2O_7 + CO_2$$
$$4\,Ac \quad + \quad 1\,Cc \quad = \quad 5\,Nfm \quad + \text{expelled } CO_2 .$$

These equations illustrate the processes of assimilation of limestone or of igneous carbonatites by basic magmas, and also the crystallization of melilite from magmas rich in carbon dioxide.

In textbooks melilites are often considered to consist essentially of the mixed crystals åkermanite and gehlenite. This statement is correct for melilites in contact-metamorphic limestones, but not for magmatic melilites, which contain essentially åkermanite and soda-melilite, while the gehlenite component is absent or quite subordinate. In fact, the average composition of magmatic melilites corresponds roughly to the formula

$$Na_{0.31}Ca_{1.71}(Mg,Fe)_{0.72}Al_{0.31}Si_{1.95}$$

which, expressed as the saturated norm, yields (putting Wo = 1)

$$Ab_{0.45}\ Wo_{1.00}\ Hy_{0.42}\ \Delta Q_{-0.41} .$$

Actual deviations from this composition are slight. In some instances, small amounts of the gehlenite component (less than 10 mol-%) or greater amounts of soda-ferrimelilite may be present according to the available amounts of anorthite or acmite. Anorthite, either contained as An' in the clinopyroxene or left over from the pyroxene calculation, can be converted into melilite, whereby some ilmenite and olivine will be involved according to the following reaction:

$$20\ CaAl_2Si_2O_8 + 4\ FeTiO_3 + (Mg,Fe)_2SiO_4 - 27\ SiO_2$$
$$= 12\ CaAl_2SiO_6 + 2\ Ca_2Al_2SiO_7 + 4\ CaTiO_3 + 6\ (Mg,Fe)Al_2O_4, \text{ i.e.}$$
$$100\ An + 8\ Il + 3\ Ol - 27\ Q = 48\ Ts + 10\ Geh + 8\ Psk + 18\ Hc\ .$$

The gehlenite will be added to the melilite, whereas the Ts is added to the remaining clinopyroxene as long as it does not exceed the amount of wollastonite, in which case the excess of Ts $(= Ts - Wo)$ will be converted into gehlenite and ilmenite:

$$CaAl_2SiO_6 + CaTiO_3 + FeO = Ca_2Al_2SiO_7 + FeTiO_3$$
$$4\ Ts\quad +\ 2\ Psk\ +\ \text{“Mt”} =\quad 5\ Geh\quad + 2\ Il$$

or, if the amount of perovskite is insufficient, according to the reaction

$$2\ CaAl_2SiO_6 + Fe_3O_4 = Ca_2Al_2SiO_7 + FeAl_2O_4 + Fe_2SiO_4\ (+O)$$
$$8\ Ts\quad + 3\ Mt =\quad 5\ Geh\quad +\quad 3\ Hc\ + 3\ Fa\ (in\ Ol)\ .$$

If, instead of anorthite, the clinopyroxene contains acmite, the desilication may result in the formation of soda-ferrimelilite: $5\ Nfm = NaCaFe^{3+}Si_2O_7$. The conversion of acmite into this melilite follows the reaction:

$$4\ NaFeSi_2O_6 + 2\ Ca_2MgSi_2O_7 - 3\ SiO_2 = 4\ NaCaFeSi_2O_7 + Mg_2SiO_4$$
$$16\ Ac\quad +\quad 10\ Åk\quad -\ 3\ Q =\quad 20\ Nfm\quad + 3\ Fo\ .$$

Because åkermanite and soda-ferrimelilite represent components of the melilite, this equation can be reduced to:

$$16\ Ac - 3\ Q = 10\ Mel + 3\ Ol\ .$$

This relation is valid only if the amount of acmite is smaller than that of wollastonite.

V. Use of the Key Tables

The following Key Tables may be used for the calculation of the stable mineral assemblages of igneous rocks in different facies. Calculating by slide rule, only integer numbers of atoms will be considered.

There are six Key Tables to be used in the calculation:

Key 1: the number of atoms and the saturated norms.

Key 2: the distribution of the atoms among stable minerals of the volcanic facies.

Key 3: idem of the "wet" subvolcanic-plutonic facies.

Key 4: idem of the "dry" subvolcanic-plutonic facies.

Key 5: idem of carbonatites.

Key 6: volume percents of the minerals, systematic position and name of the rock.

The Key Tables consist of numbered steps among which several types may be distinguished.

1. Steps Indicating Alternatives

The distinction of alternatives yields the number of the next step or that of the nomogram to be used as illustrated in the following examples:

Key 2, step 1:

① Distinguish four alternatives:

    a) the saturated norm contains *Sil* ⟶ ②

    b) the saturated norm contains *An* and *Wo* ⟶ ⑭

    c) the saturated norm contains *Ac* ⟶ ⑱

    d) the saturated norm contains *Ac* and *Ns* ⟶ ㉑.

In each of the alternatives, go next to the step indicated by the arrow.

In some cases several types of alternatives are combined and presented in the form of a tabular scheme, as e.g.

Key 2, step 14:

⑭ Determine the number of the following step according to the scheme:

		$(Or + Ab) > (Wo + Hy)$	$(Or + Ab) \leq (Wo + Hy)$
Wo > Hy	$\Delta Q \geq -100$	step ⑮	
	$\Delta Q < -100$	step ⑯	step ⑰
Wo ≤ Hy	$\Delta Q \gtrless 0$	step ㉒	

2. Steps Indicating the Rules for Calculation

Most rules of calculation are given in the form of schemes in which the first line indicates the amounts of available compounds involved. The second line indicates the fractions of these compounds entering the mineral; the sum of them yields the number of atoms contained in that mineral. The third line shows the amounts of the compounds left over after the formation of the minerals.

The unit of calculation will always be the number of atoms of that compound whose available amount is smallest. This may be illustrated by the simple example of olivine formation according to the relationship

$$4 \text{ Hy} = 3 \text{ Ol} + 1 \text{ Q} .$$

If the silica deficiency $- \Delta Q$ is less than 0.25 Hy, then $|\Delta Q|$ will be the unit of calculation, and the scheme will be as follows:

Hy	ΔQ		= available compounds
$-4|\Delta Q|$	$+|\mathbf{\Delta Q}|$	olivine $= 3|\Delta Q|$	= entering olivine
Hy*	$-$		= left over.

If, on the contrary, Hy is less than $4|\Delta Q|$, then the scheme writes:

Hy	ΔQ		= available compounds
$-\mathbf{Hy}$	$+0.25$ Hy	olivine $= 0.75$ Hy	= entering olivine
$-$	$\Delta Q*$		= left over.

In such schemes, the compounds serving as units are indicated by bold print.

In many steps several minerals will be formed, each having its own unit of calculation corresponding to the smallest amount of a remaining compound. If one of the available compounds enters totally in one of the minerals to be calculated, then the scheme will be very simple. This case is illustrated by step 3 in Key 2, concerning the calculation of cordierite, hypersthene and quartz. The total amount of Sil enters cordierite which contains furthermore some Hy, Or, Ab and ΔQ indicated by fractions of Sil which serves as a unit of calculation. The amounts of Hy* and $\Delta Q*$ remaining after the formation of cordierite correspond to the amounts of hypersthene and quartz to be formed. The relative scheme reads as follows:

③	Or	Ab	Sil	Hy	ΔQ	
	-0.02	-0.10	$-Sil$	-0.75	-0.10	*cordierite* $=$ sum
				$-Hy^*$		*hypersthene* $=Hy^*$
					$-\Delta Q^*$	*quartz* $=\Delta Q^*$
	Or^*	Ab^*	—	—	—	$\longrightarrow$ ㊳

The remaining feldspar compounds will be calculated according to the rules given in step 38 as indicated by the arrow. If, on the contrary, parts of all available compounds enter different minerals, then a value X must be calculated to serve as a unit for the calculation of one of these minerals. The scheme of step 32 in Key 2 may illustrate such a case:

㉜	Calculate: $X = q_N(Or^* + Ab^*)$			
	Or^*	Ab^*	ΔQ_1	
	$-0.83\,k^*X$	$-(2.5-0.83\,k^*)X$	$+X$	nepheline $= 1.5\,X$
	$-4.55\,|\Delta Q^*|$	$-0.45\,|\Delta Q^*|$	$+|\Delta Q^*|$	leucite $= 4\,|\Delta Q^*|$
	Or^{**}	Ab^{**}	—	$\longrightarrow$ ㊳

The values of $k^* = Or^*/(Or^* + Ab^*)$ and of q_N (read in Fig. 46) have been determined in previous steps.

3. Graphical Solutions

In order to avoid time-wasting calculations, graphs have been inserted in the Keys to determine either the number of the next step or numerical values to be used as factors in the following calculations. All graphs are entered using previously calculated coordinates.

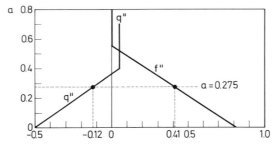

Fig. 29. $f'' - q''$ graph for alkaline amphiboles. Enter the ordinate $a = Ac/(Ac + Hy)$ into the graph and read the values of f'' and q'' on the abscissa

The nomograms indicating the next step are divided into fields. The projected point will fall in one of them, marked by the number of the next step.

In the nomograms for graphical determination of factors, the numerical values are read as abscissa of the intersection of the previously calculated ordinate with the curve or straight line representing the function (Fig. 29).

Some nomograms serve a double purpose, namely reading the number of the next step and the numerical values of certain factors, as illustrated in Figs. 30 and 31.

The feldspar nomograms (Figs. 49 and 62) represent a particular type of graphs. They serve for the determination of the amount and the nature of the feldspar or of two co-existing feldspars. The amount of feldspar components remaining after the calculation of clinopyroxene, biotite, amphibole, or feldspathoids yield the "average feldspar" $Or_xAb_yAn_z$ expressed in mol percent.

Denoting $Fsp = Or^* + Ab^* + An^*$ the molecular percentages in the average feldspar are:

$$x = 100 \ Or^*/Fsp$$
$$y = 100 \ Ab^*/Fsp$$
$$z = 100 \ An^*/Fsp$$

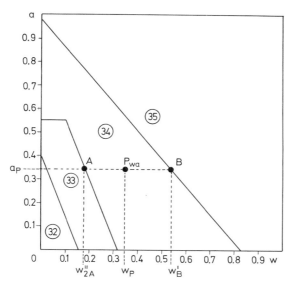

Fig. 30. $w - a$ graph for alkaline amphiboles. Enter the ordinate a_p into the graph. The point P_{wa} falls within the field ㉞ which indicates the number of the next step. Furthermore, the ordinate a_p intersects the boundary lines at A and B. The abscissa of these intersections correspond to the values of w_2'' and w', respectively

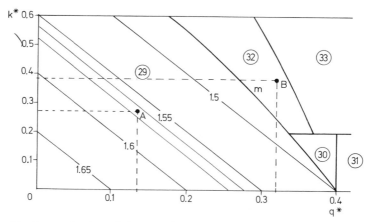

Fig. 31. Determination of the U values and the number of the next step in the $k^* - q^*$ graph for nepheline and leucite. – *Example A:* The point $A = P_{k^*q^*}$ falls within the field (29) indicating the number of the next step. Only in this field the value of U must be read by interpolation between two diagonal lines which indicate the U values. For A the value U results to be 1.565. – *Example B:* The point B falls within the field (32) indicating the number of the next step. For the determination of the q_N and q_L values see pp. 102 and 103. The equations of q_N and q_L are listened in Table 26, p. 192

rounded up to integer numbers. The numerical values of x and z determine a point P_{xz} in the feldspar nomogram. If this point falls outside the miscibility gap, then only one feldspar (sanidine, anorthoclase or plagioclase) will appear in the stable mineral assemblage. The actual composition of this feldspar will equal that of the average feldspar.

If, on the other hand, the point P_{xz} illustrating the average feldspar falls inside the gap of miscibility, then two feldspars will co-exist. The composition of these two feldspars can be deduced from the coordinates of the two end points of that tie-line which passes through the point P_{xz}. The ratio between the amounts of the two feldspars will be inversely proportional to the length of the segments between the point P_{xz} and the end points of the tie-line passing through it. In the nomograms, this ratio can be interpolated between the curves representing the amount of sanidine from ten to ten percent of the total feldspars, as shown in Fig. 32.

4. Calculation of Complex Minerals

Most femic silicate minerals contain several components of the saturated norm. For some of them, their average composition may be used for a calculation. In this case, constant ratios exist among the

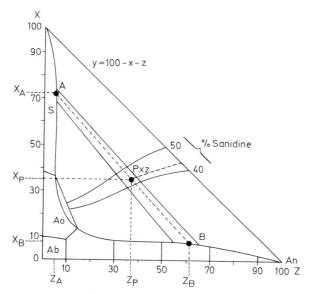

Fig. 32. Example showing the determination of the relative amount and the nature of alkaline feldspars and/or plagioclases in the volcanic facies. The point P_{xz} of the average feldspar $Or_xAb_yAn_z$ falls within the gap of miscibility on the tie-line $A---B$. The point A falls on the margin of the field of sanidine (S). Its coordinates are $x_A = 72$ and $z_A = 5$, hence $y_A = 100 - x_A - z_A = 23$, wherefrom the composition of sanidine $= Or_{72}Ab_{23}An_5$. The point B is the plot of a plagioclase with the coordinates $x_B = 8$, $z_B = 61$, hence $y_B = 31$ i.e. labradorite $= Or_8Ab_{31}An_{61}$. Interpolating along the curves, the amount of sanidine results to be 44% of the total feldspar

components entering the mineral, whatever may be the relative amounts of the corresponding components in the saturated norm of the rock. Among the available components, that which shows the relatively smallest amount will be used as a unit of calculation.

Cordierite is an example for this:

The average composition of cordierite can be expressed in the form of its saturated norm as follows:

$$100 \text{ Cord} = 51 \text{ Sil} + 38 \text{ Hy} + 5 \text{ Ab} + 1 \text{ Or} + 5 \text{ Q} .$$

The ratio between the two most important components is

$$\text{Sil} : \text{Hy} = 4 : 3 = 1.33 .$$

If in the saturated norm of the rock this ratio is greater, then Hy will be unit of calculation, i.e. all Hy will enter the cordierite and a certain amount of Sil will be left over, namely

$$Sil^* = Sil - 1.33\ Hy.$$

If, on the contrary, *Sil* is smaller than 1.33 *Hy*, then all *Sil* will enter the cordierite, and $Hy^* = Hy - 0.75\ Sil$ will be left over. Either *Hy* or *Sil* will be used as the unit of calculation according to which one of these two components is relatively the smaller.

Or	Ab	Sil	Hy	ΔQ	
0.03	0.13	1.33	**1.00**	0.14	Hy = unit
0.02	0.10	**1.00**	0.75	0.10	Sil = unit

The composition of other minerals varies with that of the magma from which they crystallized. No constant factors can be introduced into the calculation. Nepheline may be used to illustrate a case of this kind. Theoretically, nepheline should be a simple product of desilication of albite according to the relation:

$$5\ Ab - 2\ Q = 3\ Ne\ .$$

Actually, nepheline always contains some potassium and usually some silica in excess. Only in rocks which are extremely poor in silica would the ratios be $Si : Al : (Na,K) = 1 : 1 : 1$ as in the stoichiometrical formula.

In less desilicated rocks, the amount of silica in excess depends on the value of ΔQ according to the relation:

$$Ne = U \cdot |\Delta Q_1| \qquad \text{being } U \geq 1.5.$$

The numerical value of the factor U will be determined graphically. The alkali ratio k' in nepheline depends on that of

$$k^* = Or^*/(Or^* + Ab^*)$$

of the alkali feldspar compounds left over after the formation of the femic silicates. The average is $k' = 0.33\ k^*$.

On the basis of these facts, the scheme of calculation reads as follows:

Or*	Ab*	ΔQ₁	(available amounts)
$-0.33\ k^* X$	$-(1 - 0.33\ k^*)X$	$+|\Delta Q_1|$	*Nepheline* $= U \cdot |\Delta Q_1|$
Or**	Ab**	—	(remainder)

wherein $X = (1 + U) \cdot |\Delta Q_1| = $ alkaline feldspars to be converted.

The composition of clinopyroxenes is more complex, because all components of Cpx are functions of the composition of the rock, with the exception of *Wo*, which will serve as a unit of calculation. The general scheme for calculating clinopyroxenes has the following form:

Il	Or	Ab	An	Wo	Hy	ΔQ	(available amounts)
$-$ Il′	$-$ Or′	$-$ Ab′	$-$ An′	$-$ **Wo**	$-$ Hy′	$+$ Q′	*Clinopyroxene = sum*
Il*	Or*	Ab*	An*	—	Hy*	ΔQ^*	(remainder)

The determination of the various factors by which **Wo** will be multiplied to yield the numerical values of Il′, Or′, Ab′, An′, Hy′ and Q′ is made by simple calculations or by reading from nomograms, as indicated in step 22 of Key 2 or step 28 of Key 3. Analogous schemes are used for calculating alkali pyroxenes or amphiboles.

The calculation of *melilites* is a particular case which needs a few comments. Based on a series of well studied melilitites and on some melilite-bearing nephelinites and leucitites, a scheme for calculating the melilite and co-existent minerals (perovskite, hercynite, kalsilite, new olivine, etc.) was developed. It has already been stated (p. 57) that melilites are essentially "augitoids" and not feldspathoids. Their formation consists therefore in the partial or complete desilication of the previously calculated clinopyroxene.

Converting a clinopyroxene completely into melilite, olivine, perovskite and hercynite, the amount of silica $(\Sigma\,Qm)$ that becomes available to compensate the silica deficiency $(\Delta\,Qm)$ is:

a) if the Cpx contains An:

$$\Sigma\,Qm = 0.125\ \text{Wo} + 0.25\ \text{Hy}' + 0.4\ \text{Ab}' + 0.27\ \text{An}' - Q,$$

b) if the Cpx contains Ac:

$$\Sigma\,Qm = 0.125\ \text{Wo} + 0.25\ \text{Hy}' + 0.4\ \text{Ab}' + 0.187\ \text{Ac} - Q'.$$

Generally, the silica deficiency (ΔQm) left over after the formation of clinopyroxene, olivine and feldspathoids will be smaller than the silica $(\Sigma\,Qm)$ set free by the complete conversion of clinopyroxene into melilite, etc. The calculation therefore consists of two steps.

First all Cpx will be converted into melilite in order to determine the numerical value of $\Sigma\,Qm$ and the maximum amounts of melilite, etc. yielded by this operation. Multiplying these amounts by the factor

$$R = |\Delta\,Qm|/\Sigma\,Qm$$

the amounts of the newly formed minerals that are obtained are such that the silica deficiency ΔQm is exactly compensated. The rules of these calculations are given in Key 2 steps 39 to 48.

In strongly undersaturated potassic rocks with R > 0.3, kalsilite also will form at the expense of leucite:

$$4\,Lc - 1\,Q = 3\,Ks.$$

The silica set free by this reaction will compensate part of the silica deficiency (ΔQm). The factor R will be reduced to

$$R' = (|\Delta Qm| - 0.33\,Ks)/\Sigma\,Qm.$$

The amount of clinopyroxene left over after a partial conversion into melilite will be:

$$(1 - R) \cdot Cpx \quad \text{or} \quad (1 - R') \cdot Cpx, \text{ respectively}.$$

Some nepheline or leucite may be involved in the formation of melilite, as described in Key 2.

5. General Scheme of Calculation

All minerals are expressed by their saturated norms, which are to be subtracted, member by member, from the saturated norm of the rock. The remainder are distributed among other minerals, until all compounds are consumed. To accomplish these calculations the following scheme is recommended:

1st line: saturated norm of the rock,
2nd line: compounds entering the mineral A to be subtracted,
3rd line: remaining compounds (designated by asterisks in the schemes),
4th line: compounds entering the mineral B to be subtracted,
5th line: remainder (designated by double asterisks in the schemes) and so on, until no compounds are left over.

After having distributed all compounds of the saturated norm of the rock among the minerals, the sum of these minerals must equal the sum of atoms originally available, as illustrated in the examples on pp. 140–181.

Each mineral, whose amount is expressed in numbers of atoms, is to be multiplied by the specific factor given in Key 6 in order to obtain volume equivalents which, reduced to the sum 100, yield the stable mineral assemblage expressed in volume percents.

6. Determination of the Name of the Rock

In order to join the chemical composition of igneous rocks with the nomenclature, based on their mineralogical composition and the general usage among petrographers and geologists, the coordinates Q, A, P, and F must be calculated and entered in the double-triangle (Fig. 3). The projected point P_{xy} will fall in one of the numbered fields. The corresponding name of the rock is taken from Tables 19 and 20 for volcanic and for subvolcanic-plutonic rocks. Greater precision is obtained by considering the colour index and, eventually, the values of τ and σ (p. 9).

VI. Heteromorphism and Systematics

A given magma, crystallizing under different physical conditions, may yield quite different assemblages of stable minerals which, having the same bulk chemical composition, are said to be heteromorphic to each other.

1. Some Heteromorphic Igneous Rocks

In order to illustrate such heteromorphic relations, some typical examples are given in Table 15.

These few examples illustrate some very important facts concerning the nomenclature and systematics of igneous rocks and magma types. First of all, the generally accepted concept, according to which the plutonic equivalents of volcanic rocks must fall into the same field of the classificatory system, is proved to be *incorrect* in many instances. In fact, as shown in Fig. 33, *heteromorphic equivalents of igneous rocks often fall in different fields.* On the other hand, volcanic and plutonic rocks with very different chemical composition may fall into the same field, leading thus to the erroneous conclusion that they are heteromorphic to each other. Neglect of these facts very often has caused confusion in petrographical systematics, particularly with regard to undersaturated igneous rocks.

2. Definition of Plutonic Rocks

Another source of confusion is the fallacious definition of plutonic rocks, based merely on physiographic aspects, i.e. on their holocrystalline medium to coarse-grained texture. According to this definition, also granular metamorphic rocks, skarns and cumulites can be, and have been, included among plutonic rocks. In order to avoid this, the following definition of igneous plutonic rocks is here adopted:

Plutonics are rocks formed by very slow and complete crystallization of magmas under high pressure.

If the chemical composition of the magma is adequate, micas and/or amphiboles are formed under plutonic conditions. On the other hand, typical volcanic minerals, such as leucite, pigeonite, subcalcic augite and melilite, being unstable under plutonic conditions, cannot occur in true plutonites. Hence, holocrystalline granular rocks containing leucite, as

Table 15. Examples of heteromorphic rocks

Igneous rock type	Q	A	P	F	CI	Cc
1. Pantellerite	22.5	77.5	—	—	14.0	—
Soda granite	22.5	77.5	—	—	14.0	—
Riebeckite quartz-syenite	19.2	80.8	—	—	17.8	—
2. Cordierite-bearing alkali rhyolite	32.9	64.5	2.6	—	11.8	—
Two mica granite	42.2	42.8	15.0	—	15.8	—
3. Dacite	32.6	12.0	55.4	—	10.4	—
Leuco quartz-diorite	38.3	3.6	58.1	—	13.6	—
Hornblende granodiorite	33.2	13.3	53.5	—	11.8	—
4. Trachyte	0.1	80.8	19.1	—	8.3	—
Leuco augite syenite	0.2	74.0	25.8	—	8.6	—
5. Latite	0	46.4	53.6	—	18.9	—
Biotite quartz-monzodiorite	10.2	30.3	59.5	—	24.7	—
Hornblende monzonite	4.4	42.4	53.2	—	23.2	—
6. Andesite	—	—	100.0	—	36.0	—
Olivine gabbro (norite)	—	—	100.0	—	36.0	—
Hornblende quartz-gabbro	12.3	—	87.7	—	48.7	—
7. Basalt (tholeiite)	2.5	—	97.5	—	47.1	—
Gabbro (norite)	2.8	—	97.2	—	47.1	—
Hornblende quartz-gabbro	20.9	—	79.1	—	56.1	—
8. Alkali basalt	—	1.6	97.5	0.9	42.1	—
Olivine andesine gabbro	—	—	100.0	—	45.6	—
Hornblende quartz-gabbro	13.4	7.7	78.9	—	63.5	—
9. Phonolite	—	75.5	—	24.5	8.4	—
Foyaite	—	76.1	—	23.9	9.1	—
Hornblende augite foyaite	—	76.4	—	23.6	10.3	—
10. Phonolitic leucite basanite	—	11.6	47.1	41.3	27.6	—
Biotite augite plagifoyaite	—	44.8	42.2	13.0	38.6	—
Hornblende plagifoyaite	—	53.5	35.5	11.0	43.8	—
11. Nepheline basanite	—	5.0	68.0	27.0	52.1	0.4
Olivine theralite	—	2.4	72.5	25.1	54.6	0.4
Hornblendite	47.5	3.0	49.5	—	89.9	0.4
12. Melilite kalsilite nephelinite	—	—	—	100.0	59.5	2.2
Carbonatite foyaite	—	44.3	—	27.1	28.6	31.5
13. Olivine melilitite	—	—	—	100.0	84.6	1.5
Carbonatite lusitanite	—	93.6	—	6.4	57.8	31.4
14. Leucite kalsilite melilitite	—	—	—	100.0	67.2	5.3
Mafitic carbonatite ijolite	—	—	—	100.0	86.9	17.7

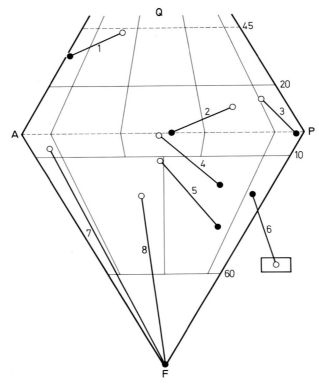

Fig. 33. Examples of heteromorphism in the STRECKEISEN double-triangle. Full circles = volcanic facies, empty circles = plutonic facies. 1 = Alkali rhyolite – Two mica granite; 2 = Latite – Biotite quartz-monzodiorite; 3 = Alkali basalt – Hornblende quartz-gabbro; 4 = Leucite phonotephrite – Biotite monzonite; 5 = Leucite phonotephrite – Biotite augite foid monzosyenite; 6 = Nepheline tephrite ("basanite") – Hornblendite; 7 = Olivine melilitite – Carbonatite lusitanite; 8 = Leucite kalsilite melilitite – Carbonatite foid monzosyenite

e.g. italite, arkite, missourite, etc., are not plutonic rocks. In fact, they are crystal aggregates accumulated at or near the top of the magma column within the vent of a volcano by gravitative or complex differentiation of the magma. Many peridotites and pyroxenites are likewise due to gravitative differentiation, by which heavy crystals accumulate, probably in the lowest part of the pyromagma, where the viscosity of the magma increases rapidly. Other coarse-granular rocks, such as diopsidic pyroxenites and uncompahgrite (essentially consisting of melilite), are better classified as skarns. Heteromorphic equivalents of such pseudoplutonics are unknown among volcanic rocks.

3. Nomenclature of Magma Types

For the above reason, magma types should not be designated by names of plutonic rocks, but by those of volcanics, the magmatic origin of which is beyond any doubt.

However, this proposal is generally not valid for carbonatite magmas. Such magmas can only exceptionally be poured out as such at the earth's surface, like the alkali carbonate magma of the volcano Oldoynio Lengai in Tanzania or the calcium carbonate magma of the small volcano Kalyanga in Uganda. Carbonatite magmas mostly lose their carbon dioxide (and water) content under low pressures and consolidate to melilitites, i.e. to extremely undersaturated calcium-rich silicate lavas which must be considered as heteromorphic equivalents of subvolcanic carbonatites.

VII. Comparison between the Stable Mineral Assemblage and the Mode

By definition, the stable mineral assemblages illustrate approximately the mineral composition attained by the rock crystallized under the conditions of the various facies and correspond to a temperature at which the crystallization was terminated. The calculation does not take into consideration possible subsolidus phase transformations and reactions, such as the formation of perthite or antiperthite in alkaline feldspars, or the unmixing of potash nepheline into nepheline and kalsilite.

On the other hand, the mode, being conditioned by the rate of cooling and degassing, may vary in a wide range. Mostly, the mode reflects a state of disequilibrium which cannot be calculated directly from the chemical bulk analysis. The stable mineral assemblage can generally not be taken as an equivalent to the actual mineral composition observable and measurable under the microscope. However, the comparison between the result of the calculation and the mode can give very useful information about the causes of discrepancies which may result, for instance, from analytical errors or omissions, incorrect mode determinations, presence of unstable phases, different degree of oxidation, and secondary alteration of the rock. These factors, liable to cause deviations of the results of the calculation from the mode are now to be discussed.

1. Analytical Errors and Omissions

Most modern rock analyses are reliable and can be used without hesitation for petrochemical calculations. On the contrary, a great number of older analyses, though reported again and again in the literature, contain erroneous determinations and must be discarded. The most frequent errors concern the values of alumina and alkalis. If they are not very small, then the calculation of the stable mineral assemblage will give results in contrast with the minerals observable under the microscope. Thus, a logical study of the stated discrepancy offers the possibility of a reciprocal control of the chemical analysis and the optical determinations.

Most chemical rock analyses found in the literature indicate the weight percentage of the most abundant cations only, expressed usually in the oxide form (SiO_2, Al_2O_3, Fe_2O_3, FeO, MnO, MgO, CaO, Na_2O, K_2O,

TiO_2, P_2O_5,$H_2O\pm$). Mostly, the contents of other elements (Cl, F, CO_2, SO_3, etc.) have not been determined. These omissions in the chemical analysis may introduce inaccuracies in the calculation and may cause important differences between the stable mineral assemblage calculated and the mode. Thus, for instance, if chlorine was not determined, the amounts of nepheline may come out too large at the expense of sodalite. If SO_3 was not given, the calculation will yield nepheline instead of haüyne.

Too great an amount of CO_2 in the analyses of undersaturated rocks will decrease the silica deficiency (ΔQ). One CO_2, determined in excess, being equivalent to one ΔQ, will produce an increase of saturated minerals at the expense of undersaturated ones. Each SiO_2, replaced by CO_2, will convert 4 Lc to 5 San, 1.5 Ne to 2.5 Ab, 3 Ol to 4 Hy, or about 5 Mel to 6 Cpx.

The omission of the determination of chlorine (and likewise of SO_3) has far reaching consequences for the calculation of minerals, as illustrated in the following stoichiometrical equations:

$$17\ NaAlSiO_4 + 2\ CaSiO_3 + (4\ Cl)$$
$$51\ Ne \qquad\quad +4\ Wo \qquad + (4\ Cl)$$
$$= 4\ Na_4Al_3Si_3O_{12}Cl + 2\ CaAl_2Si_2O_8 + NaAlSi_3O_8 + (O_2)$$
$$= 40\ Sod \qquad\qquad\quad + 10\ An \qquad\qquad + 5\ Ab$$

or also:

$$17\ KAlSi_2O_6 + 8\ NaAlSi_3O_8 + CaSiO_3 + (2\ Cl)$$
$$68\ Lc \qquad\quad + 40\ Ab \qquad\quad + 2\ Wo \ + (2\ Cl)$$
$$= 2\ Na_4Al_3Si_3O_{12}Cl + 17\ KAlSi_3O_8 + CaAl_2Si_2O_8 + (O)$$
$$= 20\ Sod \qquad\qquad\quad + 85\ Or \qquad\qquad + 5\ An\ .$$

These equations show that even a small amount of chlorine will prevent the formation of conspicuous amounts of leucite or nepheline. Furthermore, also the amounts of feldspars and even of pyroxenes (Wo) will be affected. In many cases where the determination of chlorine (or SO_3) has been omitted, the correctness of these statements can be proved by the presence of modal sodalite (or haüyne) in thin section, causing thus a striking contrast between the calculated and the modal minerals.

It is evident that the reliability of the calculation depends on the accuracy of the chemical analysis. Table 16 gives an approximate idea about the consequences of inexact chemical data. From Table 16 it is seen that particularly great errors are due to inexact determinations of the alkalis. In fact, an error of 0.1 weight percent in Na_2O causes an error of about 1 volume percent in albite, and about 0.7 volume percent in acmite. However, in good analyses the errors are much smaller, in any case smaller than those in modal analysis.

Table 16. Errors in calculating minerals and compounds caused by an error of
0.1 weight % in the chemical analysis, the sum of atoms being 1750

Error 0.1 %	Unit atom %	Mineral or compound	Symbol		Error atom %	Volume factor	Error vol.-%
SiO_2	0.095 Si	Quartz	1	Q	0.095	1.36	0.129
SiO_2	0.095 Si	Olivine	3	Ol	0.285	0.88	0.251
SiO_2	0.095 Si	Leucite	4	Lc	0.380	1.33	0.506
SiO_2	0.095 Si	Nepheline	3	Ne	0.285	1.08	0.308
SiO_2	0.095 Si	Kalsilite	3	Ks	0.285	1.17	0.333
Al_2O_3	0.112 Al	Anorthite	$2^1/_2$	An	0.280	1.21	0.339
Al_2O_3	0.112 Al	Sillimanite	$1^1/_2$	Sil	0.168	1.12	0.188
Al_2O_3	0.112 Al	Spinel	$1^1/_2$	Sp	0.168	0.88	0.148
Fe_2O_3	0.072 Fe^{3+}	Acmite	4	Ac	0.288	1.00	0.288
Fe_2O_3	0.072 Fe^{3+}	Magnetite	$1^1/_2$	Mt	0.108	0.91	0.098
FeO	0.080 Fe^{2+}	Ferrosilite	2	Fs	0.160	1.00	0.160
FeO	0.080 Fe^{2+}	Fayalite	$1^1/_2$	Fa	0.120	0.88	0.106
FeO	0.080 Fe^{2+}	Magnetite	3	Mt	0.240	0.91	0.218
MgO	0.142 Mg	Enstatite	2	En	0.284	1.00	0.284
MgO	0.142 Mg	Forsterite	$1^1/_2$	Fo	0.213	0.88	0.187
CaO	0.102 Ca	Wollastonite	2	Wo	0.204	1.00	0.204
CaO	0.102 Ca	Anorthite	5	An	0.510	1.21	0.616
Na_2O	0.185 Na	Albite	5	Ab	0.925	1.21	1.118
Na_2O	0.185 Na	Nepheline	3	Ne	0.555	1.08	0.600
Na_2O	0.185 Na	Acmite	4	Ac	0.740	1.00	0.740
Na_2O	0.185 Na	Sodasilite	$1^1/_2$	Ns	0.277	1.00	0.277
K_2O	0.122 K	Orthoclase	5	Or	0.610	1.28	0.780
K_2O	0.122 K	Leucite	4	Lc	0.488	1.33	0.649
K_2O	0.122 K	Kalsilite	3	Ks	0.366	1.17	0.428
TiO_2	0.072 Ti	Ilmenite	2	Il	0.144	0.93	0.134
TiO_2	0.072 Ti	Sphene	3	Tn	0.216	1.12	0.242
TiO_2	0.072 Ti	Perovskite	2	Psk	0.144	1.02	0.147
P_2O_5	0.080 P	Apatite	8	Ap	0.640	1.14	0.730
Cl	0.161 Cl	Sodalite	10	Sod	1.610	1.28	2.060
SO_3	0.079 SO_3	Haüyne	10	Hn	0.790	1.28	0.998
CO_2	0.130 CO_2	Calcite	2	Cc	0.260	1.11	0.288

2. Secondary Alteration

If the rock has undergone some secondary alteration, its chemical
analysis no longer reflects the original composition of the magma and
the calculation of the stable mineral assemblage loses its magmatological
significance. As was already discussed on p. 28 the secondary oxidation
of the rock causes some difference between the mode and the result of
the calculation, which is based on a standard value for the degree of
oxidation.

In many rocks, especially in acidic volcanics, which contain glass
in a more or less devitrified state, the early stages of alteration are not

easily detected under the microscope. A partial leaching of sodium and calcium will reveal itself in an excess of alumina indicated by the appearance of cordierite in the calculation. If cordierite does not exist in the mode, its presence in the calculation may be caused either by secondary alteration or by the presence of modal biotite or muscovite. Autopneumatolytic or deuteric alteration processes may affect the calculation in a way similar to that of a secondary alteration. Also if zeolites have been formed, the chemical bulk composition of the rock usually will be different from the original composition of the magma.

3. Quantitative Determination of the Mode

The cases in which the calculation gives a result closely approximating that of a quantitative mode determination are rare indeed. The volcanic rocks are generally not holocrystalline or, if they are, their grain size is mostly too small to allow the quantitative mode determination to be made with sufficient accuracy.

The mode is normally measured on thin sections by planimetric analysis (Rosiwal method) with an integrating stage or, more commonly, with a mechanical or electronic point counter or with a special eyepiece. Both integration and point counting are subjected to considerable errors, if they are not carried out with great care.

The result of a planimetric analysis may be less reliable if the rock is coarse-grained, or if it contains a few phenocrysts of relatively large size. The area of a normal thin section is too small to reflect the average mode. The area covered by the planimetric analysis must be at least one hundred times as large as the largest grains in order to warrant an accuracy of one percent. Therefore, a considerable number of thin sections should be measured. Even then, the result will be satisfactory only, if the phenocrysts are disseminated in a statistically homogeneous manner throughout the rock. In many lavas the phenocrysts are distributed somewhat irregularly and may be accumulated in clusters by laminar or turbulent flow. In such cases, a considerable amount of the rock should be crushed and quartered to obtain an average sample suitable for the chemical analysis, but it will hardly be possible to make a sufficient number of thin slides on several specimens of the rock to permit the determination of the average mode.

4. Masking Effect

If, on the other hand, the average grain size approaches the thickness of the rock section, then the grains mostly do not extend through the thin section from top to bottom. Grains of several constituents will lie on top

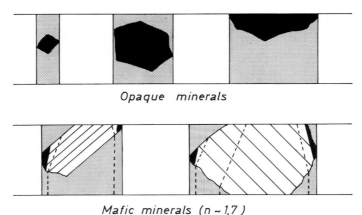

Opaque   minerals

Mafic   minerals   (n ~ 1,7)

Fig. 34. Masking effect of opaque and mafic minerals

of each other in the rock section. In such cases an opaque mineral grain prevents observation of a transparent grain above or below it. Similarly, if a high-refringent mineral grain covers a low-refringent one, it will look as if the entire depth of the rock section consists of the high-refringent mineral. These circumstances will cause an overestimation of the proportion of the opaque (magnetite, ilmenite, etc.) and high-refringent (olivine, pyroxene, etc.) minerals at the expense of the low-refringent constituents (quartz, feldspars, feldspathoids, etc.). The masking effect is schematically represented in Fig. 34 in which, in transmitted light, all the dotted area would be counted as opaque or high-refringent constituents. The errors caused by this effect will be avoided if the planimetric analysis is made in reflected light on a polished surface of the sample. The light is reflected from a virtually two-dimensional plane on which no masking effects can occur. Table 17 presents a quantitative illustration of the

Table 17. Contrast of modal determinations in thin slides and on polished sections. a = transmitted light, b = reflected light. 1 = Hawaiite, Mt. Zoccolaro (Etna, intermediate flow), 2 = Latiandesite, Mt. Zoccolaro (Etna, upper flow), 3 = Latiandesite, thick dike, Trifoglietto (Etna; after J. KLERX, 1967)

Vol.-%	1		2		3	
	a	b	a	b	a	b
Feldspars and nepheline	39.0	65.4	64.0	81.9	73.0	85.4
Augites and olivine	46.0	28.7	26.0	15.0	19.5	12.1
Opaque minerals	15.0	6.0	10.0	3.1	7.0	2.5

difference between the results of planimetric analysis carried out on three rocks in transmitted and in reflected light, respectively. As will be evident from Table 17 the errors produced by the masking effect in transmitted light may grow considerably larger than is often assumed.

The masking effect of the opaque grains increases with increasing thickness of the thin slide and with decreasing grain size of the opaque minerals. If the opaque grains are very small and the rock section rather thick, then even 10 percent by volume of magnetite may become sufficient to make the thin section look virtually opaque. If the planimetric analysis is carried out in transmitted light, the amount of the opaque mineral must be corrected for the masking effect. The correction factor to be applied can be calculated from the thickness of the rock section and from the average diameter of the opaque grains (RITTMANN and VIGHI, 1947). Fig. 35 illustrates diagrammatically the relationship between average grain size of the opaque minerals and the correction factor K, given for several thicknesses of the rock section ranging from $20-50$ μ. The validity of this method of correction has been proved recently by careful experiments (DE VECCHI, OMETTO and SCOLARI, 1968), as illustrated in Table 18 and Fig. 36.

The same correction must also be applied to the amounts of fine-grained augite and olivine obtained. Many examples of an over-estimation

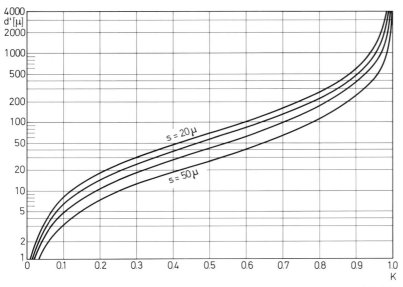

Fig. 35. Correcting factor for the masking effect. (After A. RITTMANN and L. VIGHI, 1947)

Table 18. Masking effect. (After G. DE VECCHI et al., 1968)

No. (cf. Fig. 36)	Vol.-% of the opaque minerals		Average grain size in microns
	In polished section	In thin section	
1	2.07	17.25	7
2	1.55	9.56	10
3	4.06	17.95	12
4	4.83	19.20	15
5	5.10	15.55	21
6	2.57	13.66	25
7	3.16	9.3	31
8	1.91	4.75	34
9	3.82	8.88	37
10	2.31	4.91	42

of the amounts of opaque minerals and of the femic silicate constituents can be found in the literature. In many cases the amount of magnetite given exceeds that calculable from the whole iron content as indicated by the chemical analysis. In such cases the over-estimation of the magnetite content is obviously based on a planimetric analysis in which a correction for the masking effect has been neglected.

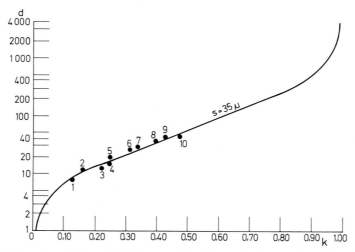

Fig. 36. Experimental proof of the correcting factor for the masking effect. (After G. DE VECCHI et al., 1968) (see Table 18)

5. Optical Determination of the Average Feldspar

The microscopic determination of the plagioclase in thin section is normally made with the universal stage. Being applicable mainly to the broad cores of the phenocrysts and not to the most narrow marginal zones, the result obtained will indicate an average composition too rich in the anorthite component.

The anorthite content of the broad and narrow zones can more readily be determined using the zonal method (RITTMANN, 1929). This method is also well suited for studying the microlitic plagioclase crystals of the fine-grained ground-mass. Also high and low temperature optics may be distinguished (RITTMANN and EL HINNAWI, 1960).

All optical methods for determining the average plagioclase composition of the entire rock are tedious. Therefore, the work is usually restricted to a determination of the extreme limits in composition between which the actual compositions of the single zones will fall. It is erroneous to take the true average anorthite content of a zoned phenocryst as equal to the arithmetic mean from the two extreme values found, or as equal to the weighted average in which the anorthite content of each zone was multiplied by the thickness of that zone. Fig. 37 illustrates the principle in calculating the average composition of a zoned plagioclase crystal. Let zone 1 represent the core, those numbered 2, 3, ... n the successive zones around the core; d_1, d_2, d_3, ... d_n the diameters of the zones (with $d_n = 1$); and An_1, An_2, An_3 .... An_n the anorthite contents of the zones. The volume fractions of the zones (V_i) will be proportional to the cube of the diameters:

$$\text{Core: } V_1 = d_1^3$$
$$V_2 = d_2^3 - d_1^3$$
$$V_3 = d_3^3 - d_2^3$$
$$\dots\dots\dots\dots\dots$$
$$V_n = 1 - d_{n-1}^3 .$$

The average anorthite content of the entire crystal section will amount to:

$$\overline{An} = \sum_1^n V_i \, An_i .$$

A rapid diagrammatic evaluation of the average anorthite content of the section of a zoned plagioclase crystal may be carried out using the graph of Figs. 38 b and c. In these, the diameter of the crystal section is taken as 1, whatever may be the direction of the chosen diameter. The anorthite content of a zone is plotted in the graph against the cube of the zone diameter (d_i^3). Connect the points indicating the various zones. A graphical integration of the curve obtained indicates the average anorthite content of that plagioclase section. Stress must be laid on the fact that the result will correspond to the average anorthite content of the entire crystal only if the studied section goes through the innermost core of the crystal.

To obtain a reliable figure for the average modal anorthite content of the plagioclase, the optical thin section study must be extended to a number of phenocrysts as well as to the microlitic crystals of the groundmass. The plagioclase microlites will usually show about the same anorthite content as the outermost

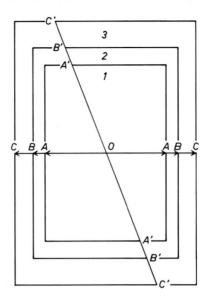

Fig. 37. Determination of the average composition of a zoned crystal. Let C—C or C'—C' be taken as unit, then the relative thicknesses A—B or A'—B' will be equal and the relative values of An% will be the same

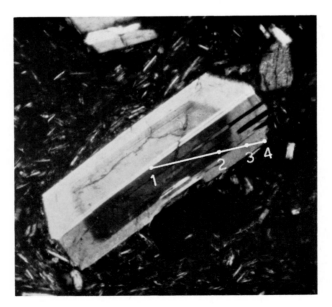

Fig. 38 a.   Zoned plagioclase in latiandesite from Filicudi (Eolian Islands). (According to L. VILLARI, personal communication)

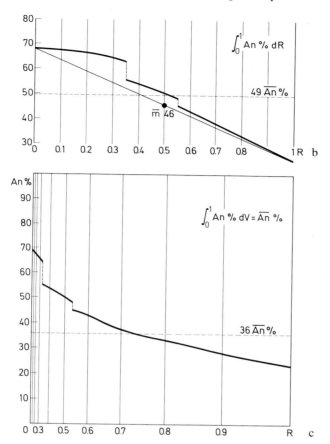

Figs. 38b and c. Determination of the average composition of the zoned crystal
of Fig. 38a

b) *Erroneous* graphical determination of the average An% by plotting the values
measured on several zones according to their relative distance from the crystal
centre along the radius R taken as unit. The curve thus obtained yields by
integration the wrong value 49 $\overline{\mathrm{An}}$%. Also the arithmetical mean results too
high with 46 $\overline{\mathrm{An}}$%.

c) *Correct* values are obtained by plotting the same values in a special graph (used
since 1942, A. RITTMANN, unpublished) as functions of the cubes of the relative
central distances. By integrating the curve thus obtained the correct value
results to be 36 $\overline{\mathrm{An}}$%

margin of the phenocrysts or will be slightly richer in albite. After the quantitative
ratio between the amount of the phenocrysts and of the microlitic plagioclase of the
groundmass has been measured, the average anorthite content of the plagioclase
in the rock may be calculated. This average plagioclase composition mostly turns
out to be richer in the albite component than would have been anticipated on the
basis of a more cursory inspection of the thin section.

In view of the difficulties encountered in an accurate determination of the average modal plagioclase composition in the rock, the calculation of the stable minerals from the chemical analysis is to be preferred. The average plagioclase composition is obtained much more rapidly by this calculation than by direct thin section study. The reliability of the calculated average plagioclase composition equals that of a mode determination in thin section.

A certain difference between the result of the calculation and the mode may arise from the fact that the amount of the potassium feldspar component (Or), incorporated in the plagioclase structure in solid solution, depends on the Ab/An ratio. Highly albitic plagioclase can take up considerably more Or than is possible for the less albitic to anorthitic plagioclases. As a consequence, a heavily zoned plagioclase crystal may contain more Or than would be possible, if the same crystal were homogeneous. Thus, in principle, zoning in plagioclase may affect the relative amounts of modal plagioclase and sanidine to some extent.

Notable differences between the relative amounts of the feldspars as revealed by the calculation and, on the other hand, as found in the quantitative mode may further arise from the temperature range of crystallization.

In internal portions of lava flows or cupolas from which the gases could not escape easily, the temperature of final crystallization may have been lower than in lavas from which the gases have escaped. The gap in solid solubility between the feldspar components will be more extended. A potassium-rich oligoclase or an anorthoclase which, according to Fig. 49 will form only one feldspar phase, may appear as a mixture of coexisting sanidine and acid plagioclase in the mode. Furthermore, interstitial sanidine or other low-refringent minerals (nepheline) in a fine-grained groundmass easily escape detection under the microscope, whereas they will be identified in the calculation.

Thus, according to a personal communication of R. ROMANO, in a series of lavas of Mount Etna, the calculation of the stable mineral association yielded 3 to 8 Vol.-% of nepheline which could not be detected under the microscope. However, their presence was confirmed by X-ray analysis.

6. Presence of Metastable Phases

The glass phase found in a volcanic rock reflects disequilibrium. The chemical composition of this metastable glass affects the bulk composition of the rock and, accordingly, will be taken into account in the calculation of the stable minerals. For such rocks the result of the calculation can only partially be compared with the mode. The difference will reveal the minerals potentially contained in the glass and, thus, may give a clue as to the true characteristics of the vitreous ground-mass.

Disequilibria in the mineral assemblage of volcanic rocks are quite commonly disclosed by the occurrence of modal crystalline constituents not appearing in the result of the calculation. So, for instance, rounded, more or less corroded olivine crystals may be present in the mode, whereas no or less olivine results from the calculation. Such a discrepancy may be understood on the basis of the incongruent melting and crystallization relationship displayed by Mg-rich pyroxenes. The olivine phenocrysts were precipitated from the melt at high temperature. On cooling these phenocrysts became unstable in the over-saturated melt. The reaction between crystals and melt was, however, not able to keep up with the rate of cooling and the olivine was partially preserved as metastable relics.

Another type of difference between the calculated stable minerals and the mode will result from the presence of intratelluric phenocrysts of constituents unstable under pure volcanic conditions. The true instability of such phenocrysts will often be evidenced in thin section by their partial replacement by stable minerals. The transformation of unstable phenocrysts into stable constituents requires favourable conditions. The rate of cooling must be low enough to allow the escape of gases at higher temperature at which the viscosity of the melt is relatively low. If the rate of cooling is high, the viscosity of the melt increases rapidly, thus preventing the escape of gases and lowering the speed of decomposing reactions.

Hydrated silicates, like biotites and amphiboles, can form only under subvolcanic or plutonic conditions, i.e. if the gas tension is high. Intratelluric phenocrysts become unstable under pure volcanic conditions at low gas tension. As soon as the gas tension of the melt is lowered beneath the vapour pressure of the hydrated silicates, they will loose water and their structures will collapse. Aggregates of new stable minerals will replace them. The following stoichiometric reaction equations illustrate the principle of the decomposition of some idealized intra-telluric phenocrysts of hydrated silicates:

$$K(Fe,Mg)_3Al\,Si_3O_{10}(OH)_2 \rightarrow KAlSi_2O_6 + (Mg,Fe)_2SiO_4 + \text{"FeO"} + H_2O$$
$$8\,Bi \qquad\qquad \rightarrow \quad 4\,Lc \quad + \quad 3\,Ol \quad + \text{"Mt"} + (aq)$$

or, in oversaturated lavas:

$$K(Mg,Fe)_3AlSi_3O_{10}(OH)_2 + 3\,SiO_2 \rightarrow KAlSi_3O_8 + 3(Mg,Fe)SiO_3 + H_2O$$
$$8\,Bi \qquad\qquad + 3\,Q \rightarrow 5\,Or \quad + \quad 6\,Cpx \quad (+aq)$$
$$Ca_2(Mg,Fe)_4Al_2Si_7O_{22}(OH)_2 \rightarrow CaMgSi_2O_6 \cdot 3(Mg,Fe)SiO_3 + CaAl_2Si_2O_8 + H_2O$$
$$15\,Ho \qquad\qquad \rightarrow \qquad 10\,Cpx \qquad + \quad 5\,An \quad (+aq).$$

In reality, these reactions are much more complicated. Interactions with other compounds existing in the melt and processes of oxidation interfere with the decomposition of biotites and amphiboles. The stable minerals resulting from the decomposition of hornblende or biotite form fine-grained aggregates which mostly preserve approximately the outlines of the transformed phenocrysts. If the rate of cooling is higher, relics of hornblende or biotite are left over within these aggregates, and with a still more rapid cooling, the phenocrysts undergo only an oxidation (oxyhornblende) and are surrounded by magnetite coronas. All these metastable intratelluric phenocrysts may not appear in the calculated stable mineral assemblage of the volcanic facies. The actual mineral composition of the rock represents a mixed facies. The phenocrysts belong to the subvolcanic or even plutonic facies, whereas the groundmass represents the pure volcanic facies (Fig. 39).

Viscosity increases exponentially with decreasing temperature, reaching very high values when the molten mass turns to glass. If, especially in acid and salic lavas, the rate of cooling is high, the escape

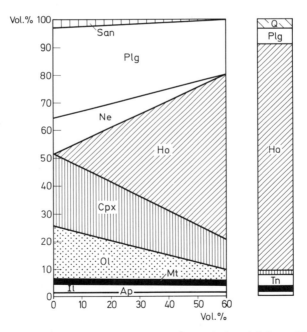

Fig. 39. Example of a mixed facies ("hornblende basalt" from Wickersberg/W.-Germany, see p. 163). Abscissa = vol.-% of hornblende phenocrysts (hastingsite), ordinate = vol.-% of the rock = mode. To the right: Mineral association of the wet plutonic facies

of volatiles will be seriously hindered. The gas tension may reach values high enough to prevent the decomposition of hydrated silicate phenocrysts. Crystallization of microlitic feldspars or other anhydrous minerals will accumulate the gases in the rest melt, increasing its gas tension. Thus, conditions may be reached, similar to those characterizing the subvolcanic or plutonic facies. Under such conditions, microlites of biotite and, less frequently, those of amphiboles will grow in the groundmass; but they may not appear in the calculation of the stable mineral assemblage.

VIII. Keys for Calculation

Key 1: Calculation of the Saturated Norm

① Multiply the weight percentage by the following factors rounding up the products to whole numbers.

SiO_2 × 16.64 = Si	ZrO_2 × 8.12 = Zr	
Al_2O_3 × 19.61 = Al	Cr_2O_3 × 13.16 = Cr	
Fe_2O_3 × 12.52 = Fe^{3+}	NiO × 13.38 = Ni	
FeO × 13.92 = Fe^{2+}	BaO × 6.52 = Ba	
MnO × 14.10 = Mn	SrO × 9.65 = Sr	
MgO × 24.80 = Mg	Li_2O × 62.93 = Li	
CaO × 17.83 = Ca	CO_2 × 22.72 = CO_2	
Na_2O × 32.27 = Na	SO_3 × 12.49 = SO_3	
K_2O × 21.23 = K	S × 31.19 = S	
TiO_2 × 12.52 = Ti	Cl × 28.20 = Cl	
P_2O_5 × 14.09 = P	F × 52.63 = F	

Add the amounts of the following atoms under simplified symbols:

$Fe^{3+} + Fe^{2+} + Mn + Cr + Ni = Fe$
$Ca + Sr = Ca;\quad K + Ba = K;\quad Mg + Li = Mg$ ⟶ ②

② Mark the degree of oxidation of the analysed rock:

$Ox^0 = Fe^{3+}/(Fe^{3+} + Fe^{2+} + Mn)$

Note that this Ox^0 will *not* be used for the calculation of magnetite (see p. 24) ⟶ ③

③ It is convenient, especially for basaltic rocks, to calculate the following values:

$$\tau = (Al_2O_3 - Na_2O)/TiO_2 \quad \text{(weight \%)}$$
$$\sigma = (Na_2O + K_2O)^2/(SiO_2 - 43) \quad \text{(weight \%)}.$$

The value τ serves to distinguish mela andesites (so-called high-alumina basalts) from other types of basalts, as well as hawaiites from andesites and mugearites from latiandesites. ⟶ ④

④ Calculation of *occasional minerals*

Calcite	Cc = 2 CO_2	remainder: Ca − CO_2
Halite	Hl = Cl	remainder: Na − Cl
Anhydrite		
a) Ca ≧ SO_3	Ah = 2 SO_3	remainder: Ca − SO_3
b) Ca < SO_3	Ah = 2 Ca	remainder: SO_3 − Ca
Thenardite	Th = 3 (SO_3 − Ca)	remainder: Na − 2 (SO_3 − Ca)
Zircon	Z = 2 Zr	remainder: Si − Zr
Pyrite	Pr = 0.5 S	remainder: Fe − 0.5 S .

Note: Occasional minerals will enter the saturated norm. In the following calculation, the remaining amounts of atoms will be designated by simple symbols, i.e. Si, Fe, Ca, Na ————————————————————→ ⑤

⑤ According to the relative amounts of atoms, five types of saturated norms will be distinguished:

a) $Al > (2\,Ca + Na + K - 3.33\,P)$ ——————————————→ ⑥

b) $(Na + K) \leqq Al \leqq (2\,Ca + Na + K - 3.33\,P)$ ——————→ ⑦

c) $Al < (Na + K) \leqq (Al + Fe^{3+})$ ————————————————→ ⑧

d) $(Na + K) > (Al + Fe^{3+})$ and $K \leqq Al$ ——————————→ ⑨

e) $K > Al$ ————————————————————————————→ ⑩

Note: Rocks containing cordierite, sillimanite, muscovite and/or garnet enter, generally, the alternative a), but weathered rocks also may show an excess in alumina. It is useless to calculate the norms of such altered rocks (see p. 49).

⑥ *Saturated norm with sillimanite*

Atoms	Ap	Il	Mt_0	Or	Ab	An	Sil	Hy	ΔQ
Si				3 K	3 Na	2 Ca*	0.5 Al*	Mg + Fe*	Si − **Si′**
Al				K	Na	2 Ca*	**Al***		
Fe		Ti	**Mt_0**					**Fe***	
Mg								**Mg**	
Ca	1.67 P					**Ca***			
Na					**Na**				
K				**K**					
Ti		**Ti**							
P	**P**								
Sum	2.67 P	2 Ti	Mt_0	5 K	5 Na	5 Ca*	1.5 Al*	2(Mg+Fe*)	Si−Si′

Enter $(Na + K)$ into Fig. 40 and read X_1 on the abscissa, then
$Mt_0 = X_1 (Fe - Ti)$,
$Si' = 3(Na + K) + 2\,Ca^* + 0.5\,Al^* + Mg + Fe$.

⑦ *Saturated norm with anorthite and wollastonite*

Atoms	Ap	Il	Mt_0	Or	Ab	An	Wo	Hy	ΔQ
Si				3 K	3 Na	Al*	Ca*	Mg + Fe*	Si − **Si′**
Al				K	Na	**Al***			
Fe		Ti	**Mt_0**					**Fe***	
Mg								**Mg**	
Ca	1.67 P					0.5 Al*	**Ca***		
Na					**Na**				
K				**K**					
Ti		**Ti**							
P	**P**								
Sum	2.67 P	2 Ti	Mt_0	5 K	5 Na	2.5 Al*	2 Ca*	2(Mg+Fe*)	Si−Si′

Enter $(Na + K)$ into Fig. 40 and read X_2 or X_3 on the abscissa:
a) if $Il \geqq 25$, then $Mt_0 = X_2 (Fe - Ti)$,
b) if $Il < 25$, then $Mt_0 = X_3 (Fe - Ti)$,
   $Si' = 3(Na + K) + Al^* + Ca^* + Mg + Fe^*$.

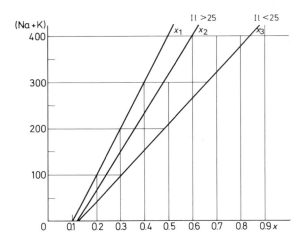

Fig. 40. Graph for the determination of the standardized magnetite factors

(8)   *Saturated norm with acmite*

Atoms	Ap	Il	Or	Ab	Ac	Mt$_0$	Wo	Hy	ΔQ
Si			3 K	3 Al*	2 Na*		Ca*	Mg + Fe*	Si − **Si'**
Al			K	**Al***					
Fe		Ti			Na*	**Mt$_0$**		**Fe***	
Mg								**Mg**	
Ca	1.67 P						**Ca***		
Na				Al*	**Na***				
K			**K**						
Ti		**Ti**							
P	**P**								
Sum	2.67 P	2 Ti	5 K	5 Al*	4 Na*	Mt$_0$	2 Ca*	2(Mg + Fe*)	Si − Si'

Enter $(Na + K)$ into Fig. 40 and read X_1 on the abscissa, then
$Mt_0 = X_1 (Fe - Ti - Na^*)$,
$Si' = 3 Al + 2 Na^* + Ca^* + Mg + Fe^*$.

⑨ *Saturated norm with acmite and sodasilite*

Atoms	Ap	Il	Or	Ab	Ac	Ns	Wo	Hy	ΔQ
Si			3 K	3 Al*	2 Fe^{3+}	0.5 Na*	Ca*	Mg + Fe*	Si − **Si'**
Al			K	**Al***					
Fe^{3+}					**Fe³⁺**				
Fe^{2+}		Ti						**Fe***	
Mg								**Mg**	
Ca	1.67 P						**Ca***		
Na				Al*	Fe^{3+}	**Na***			
K			**K**						
Ti		**Ti**							
P	**P**								
Sum	2.67 P	2 Ti	5 K	5 Al*	4 Fe^{3+}	1.5 Na*	2Ca*	2(Mg+Fe*)	Si−Si'

$Si' = 3(K + Al^*) + 2 Fe^{3+} + 0.5 Na^* + Ca^* + Mg + Fe^{2+}$.
Note: If $Ti > Fe^{2+}$, then add to Ti the amount of $Fe^{3+} = Ti − Fe^{2+}$.
Remaining Fe^{3+} to form acmite equals then $Fe^{3+} + Fe^{2+} − Ti$.

⑩ *Saturated norm with K-acmite and sodasilite*

Atoms	Ap	Il	Or	K-Ac	Ac	Ns	Wo	Hy	ΔQ
Si			3 Al	2 K*	2 Fe^{3+}*	0.5 Na*	Ca*	Mg + Fe*	Si − **Si'**
Al			**Al**						
Fe^{3+}				K*	**Fe³⁺***				
Fe^{2+}		Ti						**Fe***	
Mg								**Mg**	
Ca	1.67 P						**Ca***		
Na					Fe^{3+}*	**Na***			
K			Al	**K***					
Ti		**Ti**							
P	**P**								
Sum	2.67 P	2 Ti	5 Al	4 K*	4 Fe^{3+}*	1.5 Na*	2 Ca*	2(Mg+Fe*)	Si−Si'

$Si' = 3 Al + 2(K^* + Fe^{3+}*) + 0.5 Na^* + Ca^* + Mg + Fe^{2+}*$.

Key 2: Calculation of the Mineral Assemblage
of the Volcanic Facies

① Distinguish four alternatives[5]:

a) the saturated norm contains *Sil* ─────────────→ ②

b) the saturated norm contains *An* and *Wo* ─────────────→ ⑭

c) the saturated norm contains *Ac* and Mt_0 ─────────────→ ⑱

d) the saturated norm contains *Ac* and *Ns* ─────────────→ ㉑

---

5 If $(An + Wo) > 2(Or + Ab)$ and $ΔQ > − 300$, then it is probable that pheno-crysts have been added to the magma by gravitational differentiation. See also Keys for ultramafic rocks on page 183.

②   Calculate: $k = \text{Or}/(\text{Or} + \text{Ab})$
               $h = \text{Hy}/(\text{Sil} + \text{Hy})$
               $q = \Delta\text{Q}/(\text{Sil} + \text{Hy})$
  and distinguish two alternatives:
  a) $k < 0.30$
     enter h and q into Fig. 41 to read the number of the next step.
  b) $k \geqq 0.30$
     enter h and q into Fig. 42 to read the number of the next step.

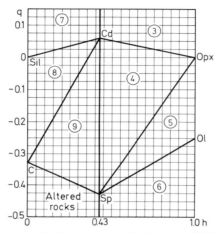

Fig. 41. $h - q$ graph for fluid magmas. Abscissa: $h = \text{Hy}/(\text{Hy} + \text{Sil})$, ordinate: $q = \Delta\text{Q}/(\text{Hy} + \text{Sil})$. Symbols see Table 21

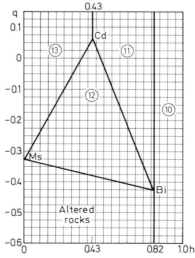

Fig. 42. $h - q$ graph for viscous magmas. Abscissa: $h = \text{Hy}/(\text{Hy} + \text{Sil})$, ordinate: $q = \Delta\text{Q}/(\text{Hy} + \text{Sil})$. Symbols see Table 21

(3)

Or	Ab	Sil	Hy	ΔQ	
-0.02	-0.10	$-$ **Sil**	-0.75 $-$ **Hy***	-0.10 $-$ **ΔQ***	*Cordierite* $=$ sum *Hypersthene* $=$ Hy* *Quartz* $=$ ΔQ*
Or*	Ab*	—	—	—	$\longrightarrow$ (38)

(4) Calculate $X = 1.23\,\Delta Q + 0.88\,\text{Sil}$

Or	Ab	Sil	Hy	ΔQ	
-0.02	-0.10	$-$ **X** $-$ **Sil***	-0.75 -0.75 $-$ **Hy***	-0.10 $+0.71$	*Cordierite* $=$ sum *Spinel* $=$ sum *Hypersthene* $=$ Hy*
Or*	Ab*	—	—	—	$\longrightarrow$ (38)

(5)

Sil	Hy	ΔQ	
$-$ **Sil**	-0.75 $-4\,\lvert\Delta Q^*\rvert$ $-$ **Hy***	$+0.71$ $+$ **ΔQ***	*Spinel* $=$ sum *Olivine* $= 3\,\lvert\Delta Q^*\rvert$ *Hypersthene* $=$ Hy*
—	—	—	$\longrightarrow$ (38)

(6)

Sil	Hy	ΔQ	
$-$ **Sil**	-0.75 $-$ **Hy***	$+0.71$ $+0.25$	*Spinel* $=$ sum *Olivine* $= 0.75$ Hy*
—	—	ΔQ_1	$\longrightarrow$ (28)

(7)

Or	Ab	Sil	Hy	ΔQ	
-0.03	-0.13	-1.33 $-$ **Sil***	$-$ **Hy**	-0.14 $-$ **ΔQ***	*Cordierite* $=$ sum *Sillimanite* $=$ Sil* *Quartz* $=$ ΔQ*
Or*	Ab*	—	—	—	$\longrightarrow$ (38)

(8)

Or	Ab	Sil	Hy	ΔQ	
-0.03	-0.13	-1.33 $-3\,\lvert\Delta Q^*\rvert$ $-$ **Sil***	$-$ **Hy**	-0.14 $+$ **ΔQ***	*Cordierite* $=$ sum *Corundum* $= 2\,\lvert\Delta Q^*\rvert$ *Sillimanite* $=$ Sil*
Or*	Ab*	—	—	—	$\longrightarrow$ (38)

⑨   Calculate: X = 0.31 Sil + 0.46 Hy + 0.93 ΔQ

Or	Ab	Sil	Hy	ΔQ	
−0.03	−0.13	−1.33	−X	−0.14	*Cordierite* = sum
		−1.33	−Hy*	+0.94	*Spinel* = sum
		−Sil*		+0.33	*Corundum* = 0.67 Sil*
Or*	Ab*	—	—	—	⟶ ㊳

⑩   Calculate: $k' = 0.3\,k + 0.70$ (see step 2)

Il	Or	Ab	Sil	Hy	ΔQ	
−0.08	−0.9 k'	−0.9(1 − k')	−Sil	−Hy	+0.52	*Biotite* = sum
Il*	Or*	Ab*	—	—	ΔQ*	

a) if ΔQ* > 0, then *Quartz* = ΔQ* ⟶ ㊳

b) if ΔQ* < 0, then ΔQ* = ΔQ₁ ⟶ ㉘

⑪   Calculate: $k' = 0.3\,k + 0.70$ (see step 2)
      X = 1.2 Hy − 0.9 Sil

Il	Or	Ab	Sil	Hy	ΔQ	
−0.08	−0.9 k'	−0.9 (1 − k')	−0.22	−X	+0.52	*Biotite* = sum
	−0.03	−0.13	−1.33	−Hy*	−0.14	*Cordierite* = sum
					−ΔQ*	*Quartz* = ΔQ*
Il*	Or*	Ab*	—	—	—	⟶ ㊳

⑫   Calculate: $k' = 0.3\,k + 0.70$ (see step 2)
      X = 0.46 Hy − 0.40 Sil − 1.21 ΔQ

Il	Or	Ab	Sil	Hy	ΔQ	
−0.08	−0.9 k'	−0.9 (1 − k')	−0.22	−X	+0.52	*Biotite* = sum
	−0.03	−0.13	−1.33	−Hy*	−0.14	*Cordierite* = sum
	−1.54	−0.31	−Sil*		+0.33	*Muscovite* = sum
Il*	Or*	Ab*	—	—	—	⟶ ㊳

⑬

Or	Ab	Sil	Hy	ΔQ	
−0.03	−0.13	−1.33	−Hy	−0.14	*Cordierite* = sum
−1.54	−0.31	−Sil*		+0.33	*Muscovite* = sum
				−ΔQ*	*Quartz* = ΔQ*
Or*	Ab*	—	—	—	⟶ ㊳

(14)   Let be $A = Or + Ab$ and $M = Wo + Hy$.
Determine the number of the following step according to the scheme:

		$A > 2\,M$	$A < 2\,M$
$Wo > Hy$	$\Delta Q \geqq -100$	step (15)	
	$\Delta Q < -100$	step (16)	step (17)
$Wo \leqq Hy$	$\Delta Q \gtrless 0$	step (22)	

(15)   Calculate: $X = 0.5\,(Wo - Hy)$ and distinguish two alternatives:

a) $Il > X$

Il	Wo	Hy	ΔQ	
$-X$	$-X$	$+X$	$-0.5\,X$	Sphene $= 1.5\,X$
Il*	Wo* $=$ Hy*		ΔQ^*	$\longrightarrow$ (22)

b) $Il \leqq X$

Il	Wo	Hy	ΔQ	
$-Il$	$-Il$	$+Il$	$-0.5\,Il$	Sphene $= 1.5\,Il$
—	Wo* $=$ Hy*		ΔQ^*	$\longrightarrow$ (22)

(16)   Calculate: $X = 0.2\,(Wo - Hy)$ and distinguish two alternatives:

a) $Il > X$

Il	An	Wo	Hy	ΔQ	
$-X$	$-X$	$-8X$	$-3X$	$+2X$	Melanite $= 11\,X$
Il*	An*	Wo* $=$ Hy*		ΔQ^*	$\longrightarrow$ (22)

b) $Il \leqq X$

Il	An	Wo	Hy	ΔQ	
$-Il$	$-Il$	$-8\,Il$	$-3\,Il$	$+2\,Il$	Melanite $= 11\,Il$
—	An*	Wo* $=$ Hy*		ΔQ^*	$\longrightarrow$ (22)

(17)   Calculate: $X = 0.5\,(Wo - Hy)$ and distinguish two alternatives:

a) $Il > X$

Il	Wo	Hy	
$-X$	$-X$	$+X$	Perovskite $= X$
Il*	Wo*	Hy*	$\longrightarrow$ (22)

b) $Il \leqq X$

Il	Wo	Hy	
$-Il$	$-Il$	$+Il$	Perovskite $= Il$
—	Wo*	Hy*	$\longrightarrow$ (22)

⑱   Distinguish two alternatives:
a) $\Delta Q \geqq 0$ ——————————————————————→ ⑲
b) $\Delta Q < 0$ ——————————————————————→ ㉓

⑲   Distinguish two alternatives:
a) 5 Mg > total Fe ——————————————————→ ㉓
b) 5 Mg ≦ total Fe ——————————————————→ ⑳

⑳   Calculate: Fs = Hy − 2 Mg (see Key 1)

Fs	ΔQ	
− Fs	+0.25	*Fayalite* = 0.75 Fs
—	ΔQ^*	——————————————→ ㉓

㉑   Distinguish two alternatives:
a) 4 (Ac + Wo + Hy) ≧ (Or + Ab) ————————————————→ ㉓
b) 4 (Ac + Wo + Hy) < (Or + Ab)
distinguish again four alternatives according to the scheme

Il/Fs =	0	0.29	0.67	greater
Ns ≧ 0.43 Fs	*A*		*B*	
Ns < 0.43 Fs		*C*	*D*	

A)

Il	Ns	Fs	ΔQ	
−Il	−1.5	−3.5	−0.5	*Cossyrite* = sum
—	Ns*	Fs*	ΔQ^*	——————→ ㉓

B)

Il	Ns	Fs	ΔQ	
−0.29	−0.43	**− Fs**	−0.15	*Cossyrite* = sum
Il*	Ns*	—	ΔQ^*	——————→ ㉓

C)

Il	Ns	Fs	ΔQ	
−Il	−1.5	−3.5 **− Fs***	−0.5 +0.25	*Cossyrite* = sum *Fayalite* = 0.75 Fs*
—	Ns*	—	ΔQ^*	——————→ ㉓

D)

Il	Ns	Fs	ΔQ	
−0.67	**− Ns**	−2.33	−0.33	*Cossyrite* = sum
Il*	—	Fs*	ΔQ^*	——————→ ㉓

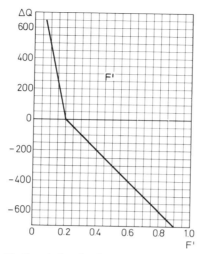

Fig. 43. Graph for the determination of the F' factor

㉒ Enter ΔQ into Fig. 43 and read F' on the abscissa:
If $\Delta Q > 0$, then $F' = 0.2\,(1 - \Delta Q \cdot 10^{-3})$,
If $\Delta Q < 0$, then $F' = 0.2 - \Delta Q \cdot 10^3$.
Calculate $F' \cdot$ Wo = maximum amount of feldspar components (Fsp')
entering clinopyroxene (Cpx). Determine the relative amounts of Or', Ab'
and An' according to the proportionality schemes:

$\Delta Q \geqq 0$	$\Delta Q < 0$
0.3 Or $= x$	0.2 Or $= x$
1.0 Ab $= y$	0.6 Ab $= y$
2.0 An $= z$	2.0 An $= z$
$x + y + z = S$	$x + y + z = S$

In both cases proceed as follows:
Calculate $D = F' \cdot$ Wo$/S$,
Or' $= x \cdot D$, Ab' $= y \cdot D$ and An' $= z \cdot D$ noting, however, that the upper limits
are Or' = Or, Ab' = Ab and An' = An.
The silica set free will be: $Q' = 0.35($Or' $+$ Ab' $+$ An'$)$.
Calculate $k =$ Or$/($Or $+$ Ab$)$ and the value of Wo/An.
Enter into Fig. 56 and read the value of p on the abscissa according to the
alternative $k < 0.4$ or $k > 0.4$. Then Il' $= p \cdot$ Il (upper limit $p = 1$).
Enter ΔQ into Fig. 44 and read the value of n on the abscissa with regard
of k. Hy' $= n \cdot$ Wo or, if Hy $< n \cdot$ Wo, then Hy' = Hy. In the latter case, no
Hy* will be left over.

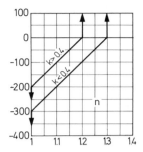

Fig. 44. Graph for the determination of the n factor. Ordinate: ΔQ

Enter the components of clinopyroxene in the following scheme:

Il	Or	Ab	An	Wo	Hy	ΔQ	
$-$Il$'$	$-$Or$'$	$-$Ab$'$	$-$An$'$	$-$Wo	$-$Hy$'$	$+$Q$'$	*provisional Cpx* = sum
Il*	Or*	Ab*	An*	—	Hy*	ΔQ*	⟶ (24)

(23) Let $M = (Ac + Wo + Hy, resp. Hy*)$ and calculate:

$Il' = 0.015\ M$ (upper limit: $Il' = Il$)
$Or' = 0.015\ M$; $Ab' = 0.125\ M$; $Q' = 0.005\ M$.

Enter these values into the following scheme:

Il	Or	Ab	Ac	Wo	Hy	ΔQ	
$-$Il$'$	$-$Or$'$	$-$Ab$'$	$-$Ac	$-$Wo	$-$Hy	$+$Q$'$	*Clinopyroxene* = sum
Il*	Or*	Ab*	—	—	—	ΔQ*	⟶ (26)

Calculate $a = Ac/M$ and distinguish the following types of alkaline clinopyroxenes:

$a \geq 0.70 = Aegirine$,
$0.70 > a > 0.15 = Aegirine\text{-}augite,$
$0.15 \geq a > 0 = Sodian\ augite.$

(24) Distinguish two alternatives:
    a) $k < 0.30$ ———————————————————————⟶ (26)

b) $k \geq 0.30$. Calculate $v = (Or + Ab + An)/(Wo + Hy)$.
Enter k and v into Fig. 45 to read the number of the next step. If the projecting point falls near to B_1 or B_2, then calculate both variants, with or without biotite. Generally, the mode will be in between them.

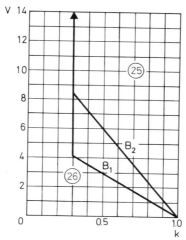

Fig. 45. Biotites in the $k - v$ diagram. Abscissa: $k = Or/(Or + Ab)$, ordinate: $v = (Or + Ab + An)/(Wo + Hy)$. B_1 if $\Delta Q > 0$, B_2 if $\Delta Q \leq 0$

(25)   Calculate $X = 0.96\ Hy^*$ and $k' = 0.3\ k + 0.70$. Distinguish two alternatives:
a) $(Or^* + Ab^*) > 1.11\ X$

Il*	Or*	Ab*	An*	Wo* = 0	Hy*	ΔQ*
−0.08	−0.9 k'	−0.9(1 − k')	−0.10	+0.04 − Wo**	− X − Hy**	+0.52 *Biotite* enter Cpx
Il**	Or**	Ab**	An**	—	—	ΔQ** ⟶ (26)

b) $(Or^* + Ab^*) \leq 1.11\ X$. Calculate: $Y = (1.2k - 0.16) \cdot (Or^* + Ab^*)$.

Il*	Or*	Ab*	An*	Wo* = 0	Hy*	ΔQ*
−0.09	−k'Y	−(1 − k')Y	−0.11	+0.04 − Wo**	− 1.11 − 0.06Hy*	+0.58 *Biotite* = sum enter Cpx
Il**	Or**	Ab**	An**	—	—	ΔQ** ⟶ (26)

Note: If $An^* < 0.1\ Hy^*$, then the above schemes cannot be used, and must be replaced by the following calculation:
Let be $X = Hy^* - 0.4\ An^*$ and use the following scheme:

Il*	Or*	Ab*	An*	Wo* = 0	Hy*	ΔQ*
−0.06	−0.9 k'	−0.9(1 − k')	− An*	+0.04 An* − Wo**	− X − Hy**	+0.52 *Biotite* = sum enter Cpx
Il**	Or**	Ab**	—	—	—	ΔQ** ⟶ (26)

The definite composition of the provisional clinopyroxene has been obtained by the addition of occasional Wo** and Hy** (or 0.06 Hy*) left over after the formation of biotite. The following types can be distinguished:

Calculate $c = (Wo' + 0.4\,An')/Hy'$ and split Hy' into Fs and En (En = 2 Mg).

c'	$En > Fs$	$En < Fs$
> 0.9	*diopside*	*hedenbergite*
0.5 to 0.9	*augite*	*ferroaugite*
0.3 to 0.5	*subcalcic augite*	*subcalcic ferroaugite*
0.1 to 0.3	*pigeonite*	*ferropigeonite.*

Augites and diopsidic augites ($c' > 0.5$) containing more than 4 % Il in their norm (atoms) are called *titanaugites*.

Note that the ratio En/Fs increases with increasing oxidation. The above subdivision of calculated clinopyroxenes refers to the degree of oxidation resulting from the calculation of standard magnetite (Mt_0).

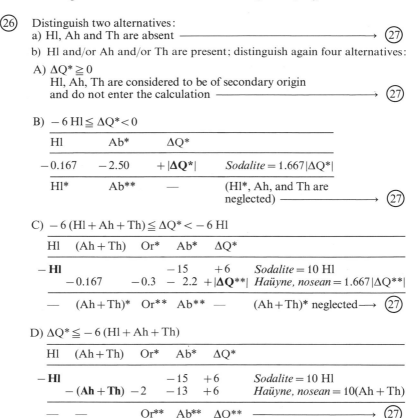

(26) Distinguish two alternatives:
a) Hl, Ah and Th are absent ————————————————→ (27)

b) Hl and/or Ah and/or Th are present; distinguish again four alternatives:

A) $\Delta Q^* \geqq 0$
Hl, Ah, Th are considered to be of secondary origin
and do not enter the calculation ————————————→ (27)

B) $-6\,Hl \leqq \Delta Q^* < 0$

Hl	Ab*	ΔQ^*	
−0.167	−2.50	$+|\Delta Q^*|$	*Sodalite* $= 1.667|\Delta Q^*|$
Hl*	Ab**	—	(Hl*, Ah, and Th are neglected) ————→ (27)

C) $-6\,(Hl + Ah + Th) \leqq \Delta Q^* < -6\,Hl$

Hl	(Ah + Th)	Or*	Ab*	ΔQ^*	
−Hl		−15	+6		*Sodalite* = 10 Hl
	−0.167	−0.3	−2.2	$+|\Delta Q^{**}|$	*Haüyne, nosean* $= 1.667|\Delta Q^{**}|$
—	(Ah + Th)*	Or**	Ab**	—	(Ah + Th)* neglected ——→ (27)

D) $\Delta Q^* \leqq -6\,(Hl + Ah + Th)$

Hl	(Ah + Th)	Or*	Ab*	ΔQ^*	
−Hl		−15	+6		*Sodalite* = 10 Hl
	−(Ah + Th) −2	−13	+6		*Haüyne, nosean* = 10(Ah + Th)
—	—	Or**	Ab**	ΔQ^{**}	————————→ (27)

(27) Distinguish three alternatives according to the relative values of ΔQ^* (resp. ΔQ^{**}). Note that Hy* (resp. Hy) may be nil.

a) $\Delta O^* > 0$:
  $\Delta Q^* = Quartz$
  Hy* = Hypersthene ————————————————————→ (38)

b) $-0.25\,\text{Hy}^* \leq \Delta Q^* < 0$:
  $3|\Delta Q^*| = Olivine$
  Hy* $- 4|\Delta Q^*|$ = Hypersthene ———————————→ (38)

c) $\Delta Q^* < -0.25\,\text{Hy}^*$:
  $0.75\,\text{Hy}^* = Olivine$
  $\Delta Q^* + 0.25\,\text{Hy}^* = \Delta Q_1$ ———————————→ (28)

After the formation of olivine, the definite composition of clinopyroxene is known and permits the determination of the type according to the calculation shown in step 25.

(28) Calculate: $k = Or/(Or + Ab)$; $k^* = Or^*/(Or^* + Ab^*)$
      and $q^* = |\Delta Q^*|/(Or^* + Ab^*)$.
To establish the number of the next step distinguish 3 alternatives:

a) $k < 0.4$
   enter k^* and q^* into Fig. 46.

b) $0.4 \leq k \leq 0.58$
   Calculate $v' = (Or + Ab)/(Wo + Hy)$;
   enter k and v' into Fig. 47 to establish which alternative (28a or 28c) must be used for entering k^* and q^*.

c) $k > 0.58$
   enter k^* and q^* into Fig. 48.

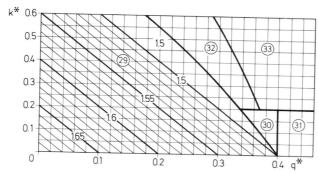

Fig. 46. $q^* - k^*$ graph for fluid magmas for the calculation of leucite and nepheline. Abscissa: $q^* = |\Delta Q^*|/(Or^* + Ab^*)$, ordinate: $k^* = Or^*/(Or^* + Ab^*)$. *Determination of q_N:* enter the ordinate k^* into the graph, but only till to the curve which separates field 29 from the fields 32 and 30 and read the value of q_N on the abscissa

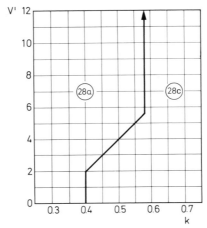

Fig. 47. $k - v'$ graph. Abscissa: $k = Or/(Or + Ab)$, ordinate: $v' = (Or + Ab)/(Wo + Hy)$

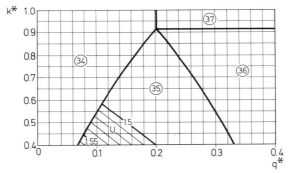

Fig. 48. $q* - k*$ graph for viscous magmas for the calculation of nepheline and leucite. Abscissa: $q* = |\Delta Q*|/(Or* + Ab*)$, ordinate: $k* = Or*/(Or* + Ab*)$. *Determination of q_L: enter the ordinate $k*$ into the graph, but only till to the curve which separates field 34 from 35 and read the value of q_L on the abscissa. Upper limit of $q_L = 0.2$*

㉙ Read in Fig. 46 the value of U corresponding to P_{k*q*}
     and calculate $X = (1 + U)|\Delta Q_1|$

Or*	Ab*	ΔQ_1	
$-0.33\ k*\,X$	$-(1 - 0.33\ k*)X$	$+|\Delta Q_1|$	*Nepheline* $= U \cdot |\Delta Q_1|$
Or**	Ab**	—	⟶ ㊳

③⓪

(Or* + Ab*)	ΔQ_1	
$-2.5|\Delta Q_1|$	$+|\Delta Q_1|$	*Nepheline* $= 1.5\ |\Delta Q_1|$
(Or** + Ab**)	—	
being Or** $= 3.5\ k^*$(Or** + Ab**)		$\longrightarrow$ ③⑧

③①

Or*	Ab*	ΔQ_1	
$-$Or*	$-$Ab*	$+0.4$(Or* + Ab*)	*Nepheline* $= 0.6$(Or* + Ab*)
—	—	ΔQ_m	$\longrightarrow$ ④⓪

③② Read in Fig. 46 on the abscissa the value of q_N corresponding to k^* and calculate: $X = q_N$(Or* + Ab*)

Or*	Ab*	ΔQ_1	
$(-0.83\ k^*)X$ $-4.55|\Delta Q_1^*|$	$-(2.5-0.83\ k^*)X$ $-0.45|\Delta Q_1^*|$	$+X$ $+|\Delta Q_1^*|$	*Nepheline* $= 1.5\ X$ *Leucite* $= 4|\Delta Q_1^*|$
Or**	Ab**	—	$\longrightarrow$ ③⑧

③③ Read in Fig. 46 on the abscissa the value of q_N corresponding to k^* and calculate: $X = q_N$(Or* + Ab*)

(Or* + Ab*)	ΔQ_1	
$-2.5\,X$ $-$(Or** + Ab**)	$+X$ $+0.2$(Or** + Ab**)	*Nepheline* $= 1.5\ X$ *Leucite* $= 0.8$(Or** + Ab**)
—	ΔQ_m	$\longrightarrow$ ④⓪

③④

Or*	Ab*	ΔQ_1	
$-4.55|\Delta Q_1|$	$-0.45|\Delta Q_1|$	$+|\Delta Q_1|$	*Leucite* $= 4|\Delta Q_1|$
Or**	Ab**	—	$\longrightarrow$ ③⑧

③⑤ Read in Fig. 48 on the abscissa the value of q_L corresponding to k^* and calculate: $X = q_L$(Or* + Ab*)

Or*	Ab*	ΔQ_1	
$-4.55\,X$ $-0.83\ k^*|\Delta Q_1^*|$	$-0.45\,X$ $-(2.5-0.83\ k^*)|\Delta Q_1^*|$	$+X$ $+|\Delta Q_1^*|$	*Leucite* $= 4\ X$ *Nepheline* $= U|\Delta Q_1^*|$
Or**	Ab**	—	$\longrightarrow$ ③⑧

㊱   Read in Fig. 48 on the abscissa the value of q_L corresponding to k^* and
    calculate: $X = q_L(Or^* + Ab^*)$

(Or* + Ab*)	ΔQ_1	
$-5\,X$	$+X$	*Leucite* $= 4\,X$
$-(Or^{**} + Ab^{**})$	$+0.4(Or^{**} + Ab^{**})$	*Nepheline* $= 0.6(Or^{**} + Ab^{**})$
—	ΔQ_m	$\longrightarrow$ ㊵

㊲

Or*	Ab*	ΔQ_1	
$-Or^*$	$-Ab^*$	$+0.2(Or^* + Ab^*)$	*Leucite* $= 0.8(Or^* + Ab^*)$
—	—	ΔQ_m	$\longrightarrow$ ㊵

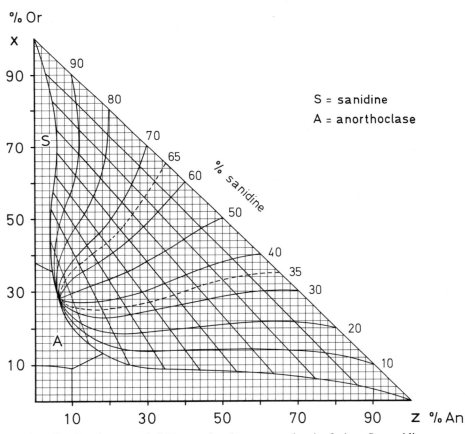

Fig. 49. Coexistence of feldspars in the pure volcanic facies. S = sanidine,
A = anorthoclase

(38) Calculate the "average feldspar" $Or_x Ab_y An_z$ on the basis of the feldspar compounds Or* (resp. Or**), Ab* (resp. Ab**) and An* left over after the formation of clinopyroxene, biotite, and feldspathoids.
Denoting Or* + Ab* + An* = Fsp* (resp. Or** + Ab** + An** = Fsp*) and determine:

$$x = 100 \ Or^*/Fsp^*$$
$$y = 100 \ Ab^*/Fsp^*$$
$$z = 100 \ An^*/Fsp^*.$$

Enter x and z into Fig. 49 to determine the nature and amount (in %) of the feldspar(s). Enter into Key 6 (p. 131).

(39) Calculate provisional clinopyroxene (Cpx_0) replacing in the already calculated Cpx the value of Il' by Il″ = 0.5 Il and that of An' by (An' + An*). The other half of Il, i.e. Il″ is added to magnetite (Mt_0).
Convert this Cpx_0 completely into provisional melilite (Mel_0), olivine (Ol_0), hercynite (Hz_0), perovskite (Psk_0) and nepheline (Ne_0). Let ΣQ_m be the maximum amount of silica set free by this conversion. ────→ (40)

(40) Distinguish three alternatives:

a) An > 2.5 Il″ ──────────────────────────────→ (41)

b) An ≤ 2.5 Il″ ──────────────────────────────→ (42)

c) An = 0 (Ac is present) ──────────────────────→ (43)

---

(41)

Il″	(Or' + Ab')	An	Wo	Hy' − Q'	= Cpx_0
− Il″		− 2.5		+ 1.00	Psk_0 = Il″; Hz_0 = 1.5 Il″
		− An* − 0.4		+ 0.40 ⎱	
	− 0.45		− Wo* − 0.42	+ 0.41 ⎰	Mel_0 = sum
				− Hy* + 0.25	Ol_0 = 0.75 Hy*
	− (Or' + Ab')*			+ 0.40	Ne_0 = 0.6(Or' + Ab')*
— —		—	—	−	
				ΣQ_m	──────────→ (44)

---

(42)

Il″	(Or' + Ab')	An	Wo	Hy' − Q'	= Cpx_0
− 0.4		− An		+ 0.40	Hz_0 = 0.6 An
− Il*			− 1.0 + 1.0		Psk_0 = Il″
	− 0.45		− Wo* − 0.42	+ 0.41	Mel_0 = sum
				− Hy* + 0.25	Ol_0 = 0.75 Hy*
	− (Or' + Ab')*			+ 0.40	Ne_0 = 0.6(Or' + Ab')*
— —		—	—	−	
				ΣQ_m	──────────→ (44)

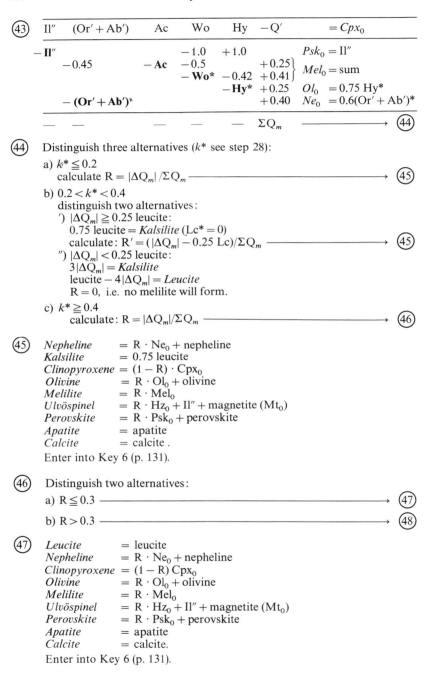

㊸ Il'' $(Or' + Ab')$ Ac Wo Hy $-Q'$ $= Cpx_0$

$-\mathbf{Il''}$ -1.0 $+1.0$ $Psk_0 = Il''$
       -0.45 $-\mathbf{Ac}$ -0.5 $+0.25\}$ $Mel_0 = \text{sum}$
                                 $-\mathbf{Wo^*}$ -0.42 $+0.41\}$
                                                $-\mathbf{Hy^*}$ $+0.25$ $Ol_0 = 0.75\ Hy^*$
       $-(\mathbf{Or' + Ab'})^?$ $+0.40$ $Ne_0 = 0.6(Or' + Ab')^*$

$-$ $-$ $-$ $-$ $-$ ΣQ_m ⟶ ㊹

㊹ Distinguish three alternatives (k^* see step 28):

   a) $k^* \leq 0.2$
      calculate R $= |\Delta Q_m| / \Sigma Q_m$ ⟶ ㊹

   b) $0.2 < k^* < 0.4$
      distinguish two alternatives:
      ') $|\Delta Q_m| \geq 0.25$ leucite:
         0.75 leucite $= Kalsilite$ (Lc* = 0)
         calculate: R' $= (|\Delta Q_m| - 0.25\ Lc)/\Sigma Q_m$ ⟶ ㊹
      ") $|\Delta Q_m| < 0.25$ leucite:
         $3|\Delta Q_m| = Kalsilite$
         leucite $- 4|\Delta Q_m| = Leucite$
         R = 0, i.e. no melilite will form.

   c) $k^* \geq 0.4$
      calculate: R $= |\Delta Q_m|/\Sigma Q_m$ ⟶ ㊺

㊹ *Nepheline* $= R \cdot Ne_0 + \text{nepheline}$
   Kalsilite $= 0.75$ leucite
   Clinopyroxene $= (1 - R) \cdot Cpx_0$
   Olivine $= R \cdot Ol_0 + \text{olivine}$
   Melilite $= R \cdot Mel_0$
   Ulvöspinel $= R \cdot Hz_0 + Il'' + \text{magnetite} (Mt_0)$
   Perovskite $= R \cdot Psk_0 + \text{perovskite}$
   Apatite $= \text{apatite}$
   Calcite $= \text{calcite}$.
   Enter into Key 6 (p. 131).

㊺ Distinguish two alternatives:

   a) R ≤ 0.3 ⟶ ㊼

   b) R > 0.3 ⟶ ㊽

㊼ *Leucite* $= \text{leucite}$
   Nepheline $= R \cdot Ne_0 + \text{nepheline}$
   Clinopyroxene $= (1 - R)\ Cpx_0$
   Olivine $= R \cdot Ol_0 + \text{olivine}$
   Melilite $= R \cdot Mel_0$
   Ulvöspinel $= R \cdot Hz_0 + Il'' + \text{magnetite} (Mt_0)$
   Perovskite $= R \cdot Psk_0 + \text{perovskite}$
   Apatite $= \text{apatite}$
   Calcite $= \text{calcite}$.
   Enter into Key 6 (p. 131).

(48) Calculate: $X = (R - 0.3) \cdot (|\Delta Q_m| + |Q'|)$ and distinguish two alternatives:
a) $X \leq 0.75$ leucite, then *kalsilite* = X
b) $X > 0.75$ leucite, then *kalsilite* = 0.75 leucite (Lc* = 0).
In both cases $R^* = (|\Delta Q_m| - 0.333 \text{ kalsilite})/\Sigma Q_m$.

Leucite	= leucite $- 1.333$ kalsilite
Kalsilite	= X or 0.75 leucite (see above)
Nepheline	= $R^* Ne_0 +$ nepheline
Clinopyroxene	= $(1 - R^*) Cpx_0$
Olivine	= $R^* Ol_0 +$ olivine
Melilite	= $R^* Mel_0$
Ulvöspinel	= $R^* Hz_0 + Il'' +$ magnetite (Mt_0)
Perovskite	= $R^* Psk_0 +$ perovskite
Apatite	= apatite
Calcite	= calcite .

Note: R, resp. R* may be greater than 1.0.
In this case, calculate as before, but with R = 1 (resp. R* = 1). The provisional minerals (Mel_0, Ol_0, Psk_0, Hz_0 and Ne_0) will become definite and no clinopyroxene will be left over.
The remaining silica deficiency amounts $\Delta Q_r = \Sigma Q_m - \Delta Q_m$.
Distinguish two cases:
a) *CO_2 has been determined* and calcite has been calculated:
Compensate ΔQ_r according to the following conversions:
1) if leucite is present: $4 \text{ Lc} - 1 \text{ Q} = 3 \text{ Ks}$
2) if olivine is present: $3 \text{ Fa} - 1 \text{ Q} = 2 \text{ "Mt"}$
(this concerns only the fayalite component of olivine and includes an oxidation)
3) 10 melilite (Åk) $+ 3 \text{ Ol} - 1 \text{ Q} = 12$ monticellite.
b) *CO_2 has not been determined.* Leave ΔQ_r as it is, indicating thus that the analysis is incomplete.

Enter into Key 6 (p. 131).

Key 3: Calculation of the Mineral Assemblage of the "Wet" Subvolcanic-Plutonic Facies

(1) Distinguish three alternatives[6]:
a) the saturated norm contains *Sil* ⎯⎯⎯⎯⎯⎯⎯⎯→ (2)
b) the saturated norm contains *An* and *Wo* ⎯⎯⎯⎯→ (12)
c) the saturated norm contains *Ac* ($\pm$ *Ns*) ⎯⎯⎯→ (29)

(2) Distinguish two alternatives:
a) $Fs < 3 En$ (i.e. $m > 0.25$) ⎯⎯⎯⎯⎯⎯⎯→ (3)
b) $Fs \geq 3 En$ (i.e. $m \leq 0.25$) ⎯⎯⎯⎯⎯⎯→ (8)

6 Remark: if $(Wo + Hy + Ac) > 2(Or + Ab + An)$, then use the Key for ultramafic rocks on page 183.

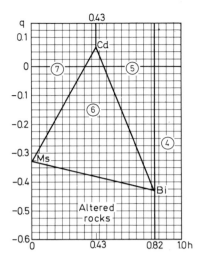

Fig. 50. $h-q$ graph with cordierite. Abscissa: $h = H/(Hy + Sil)$, ordinate: $q = \Delta Q/(Hy + Sil)$. Symbols see Table 21

③   Calculate: $h = Hy/(Sil + Hy)$ and $q = \Delta Q/(Sil + Hy)$. Enter h and q in Fig. 50 and read the number of the next step. Furthermore calculate $k = Or/(Or + Ab)$.

④   Calculate: $k' = 0.3k + 0.70$ and distinguish two alternatives:

a) $Hy \leqq 0.9(Or + Ab)$

Il	Or	Ab	Sil	Hy	ΔQ	
-0.08	$-0.9k'$	$-0.9(1-k')$	$-Sil$	$-\mathbf{Hy}$	$+0.52$	*Biotite*
Il*	Or*	Ab*	—	—	ΔQ*	

if $\Delta Q^* \geqq 0$, then $\Delta Q^* = Quartz$ ————————————————→ ㊴

if $\Delta Q^* < 0$, then $\Delta Q^* = \Delta Q_1$ ————————————————→ ㊱

b) $Hy > 0.9(Or + Ab)$

    Calculate: $X = 1.2(Or - 0.1 Ab) = $ unit

Il	Or	Ab	Sil	Hy	ΔQ	
-0.08	$-0.9k'$	$-0.9(1-k')$	-0.22	$-\mathbf{X}$	$+0.52$	*Biotite*
	-0.02	-0.10	$-\mathbf{Sil^*}$	-0.75	-0.10	*Cordierite*
Il*	Or*	Ab*	—	Hy*	ΔQ*	———→ ㉔

(5) Calculate: $X = 1.2$ Hy $- 0.9$ Sil and $k' = 0.3 k + 0.70$

Il	Or	Ab	Sil	Hy	ΔQ	
−0.08	−0.9 k'	−0.9 $(1 - k')$	−0.22	−**X**	+0.52	*Biotite*
	−0.03	−0.13	− Sil*	− **Hy***	−0.14	*Cordierite*
					+**ΔQ***	*Quartz*
Il*	Or*	Ab*	—	—	—	→ (39)

(6) Calculate: $X = 0.567$ Hy $- 0.324$ Sil $- 0.971$ ΔQ
and $k' = 0.3 k + 0.70$

Il	Or	Ab	Sil	Hy	ΔQ	
−0.08	−0.9 k'	−0.9 $(1 - k')$	−0.22	−**X**	+0.52	*Biotite*
	−0.03	−0.13	−1.33	− **Hy***	−0.14	*Cordierite*
	−1.54	−0.31	− Sil*		+0.33	*Muscovite*
Il*	Or*	Ab*	—	—	—	→ (39)

(7)

Or	Ab	Sil	Hy	ΔQ	
−0.03	−0.13	−1.33	− **Hy**	−0.14	*Cordierite*
−1.54	−0.31	− Sil*		+0.33	*Muscovite*
				+ **ΔQ***	*Quartz*
Or*	Ab*	—	—	—	→ (39)

(8) Calculate: $h = $ Hy/(Sil + Hy) and $q = $ ΔQ/(Sil + Hy). Enter h and q in Fig. 51 and read the number of the next step.

(9) Calculate: $X = 1.79$ (Hy $- 2$Sil) and $k' = 0.3 k + 0.70$

Il	Or	Ab	Sil	Hy	ΔQ	
−0.08	−0.9 k'	−0.9 $(1 - k')$	−0.22	−**X**	+0.52	*Biotite*
			−0.50	− **Hy***	+0.17	*Garnet*
					+**ΔQ***	*Quartz*
Il*	Or*	Ab*	—	—	—	→ (39)

(10) Calculate: $X = 2.24 |$ΔQ$| - 0.75$ Sil and $k' = 0.3 k + 0.70$

Il	Or	Ab	Sil	Hy	ΔQ	
−0.08	−0.9 k'	−0.9 $(1 - k')$	−0.22	−**X**	+0.52	*Biotite*
			−0.50	− **Hy***	+0.17	*Garnet*
	−1.54	−0.31	− Sil*		+**ΔQ***	*Muscovite*
Il*	Or*	Ab*	—	—	—	→ (39)

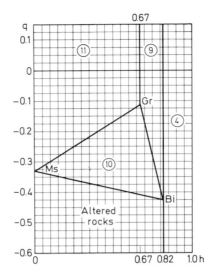

Fig. 51. $h - q$ graph including garnet. Abscissa: $h = \mathrm{Hy}/(\mathrm{Hy} + \mathrm{Sil})$, ordinate: $q = \Delta Q/(\mathrm{Hy} + \mathrm{Sil})$. Symbols see Table 21

⑪	Or	Ab	Sil	Hy	ΔQ	
			0.50	−**Hy**	+0.17	*Garnet*
	−1.54	−0.31	−**Sil***		+0.33	*Muscovite*
					+**ΔQ***	*Quartz*
	Or*	Ab*	—	—	—	⟶ ㊴

⑫ Calculate: $w = \mathrm{Wo}/\mathrm{Hy}$ and distinguish three alternatives:
  a) $0 < w < 0.5$ then calculate: $X = 2w \cdot \mathrm{Il}$
  b) $0.5 \leq w < 1.0$ then calculate: $X = 2(1 - w) \cdot \mathrm{Il}$
  In both cases (a and b), calculate sphene or melanite according to the following schemes:

') if $\Delta Q > -100$:

Il	Wo	Hy	ΔQ	
−X	−X	+X	−0.5 X	*Sphene* = 1.5 X
Il*	Wo*	Hy*	ΔQ*	⟶ ⑰

") if $\Delta Q < -100$:

Il	An	Wo	Hy	ΔQ	
$-X$	$-X$	$-8X$	$-3X$	$+2X$	*Melanite* $= 11X$
Il*	An*	Wo*	Hy*	ΔQ^*	⟶ ⑰

c) $w > 1.0$ ⟶ ⑬

⑬

	$(Or + Ab) \geqq 2(Wo + Hy)$	$(Or + Ab) < 2(Wo + Hy)$
$\Delta Q \geqq 0$	14	14
$\Delta Q < 0$	15	16

⑭ Let be $X = 0.5(Wo - Hy)$ and distinguish two alternatives:

a) $Il > X$

Il	Wo	Hy	ΔQ	
$-X$	$-X$	$+X$	$-0.5X$	*Sphene* $= 1.5X$
Il*	Wo*	$=$ Hy*	ΔQ^*	⟶ ⑰

b) $Il \leqq X$

Il	Wo	Hy	ΔQ	
$-Il$	$-Il$	$+Il$	$-0.5\,Il$	*Sphene* $= 1.5\,Il$
—	Wo*	$=$ Hy*	ΔQ^*	⟶ ⑰

⑮ Let be $X = 0.2(Wo - Hy)$ and distinguish two alternatives:

a) $Il > X$

Il	An	Wo	Hy	ΔQ	
$-X$	$-X$	$-8X$	$-3X$	$+2X$	*Melanite* $= 11X$
Il*	An*	Wo*	Hy*	ΔQ^*	⟶ ⑰

b) $Il < X$

Il	An	Wo	Hy	ΔQ	
$-Il$	$-Il$	$-8\,Il$	$-3\,Il$	$+2\,Il$	*Melanite* $= 11\,Il$
—	An*	Wo*	Hy*	ΔQ^*	⟶ ⑰

(16) Let be $X = 0.5(Wo - Hy)$ and distinguish two alternatives:
a) $Il > X$

Il	Wo	Hy	
$-X$	$-X$	$+X$	$Perovskite = X$
Il^*	Wo^*	Hy^*	

→ (17)

b) $Il < X$

Il	Wo	Hy	
$-Il$	$-Il$	$+Il$	$Perovskite = Il$
—	Wo^*	Hy^*	

→ (17)

(17) Calculate: $k = Or/(Or + Ab)$, $w^* = Wo^*/Hy^*$ and $m = En/Hy^*$.
Distinguish two alternatives:

a) $\Delta Q \geqq 0$
enter w^* and m in Fig. 52 and read the number of the next step.

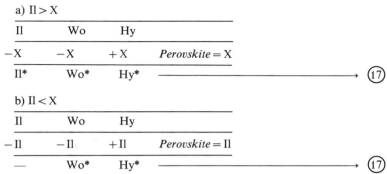

Fig. 52. $w^* - m$ graph for amphiboles in oversaturated rocks ($\Delta Q > 0$).
Abscissa: $w^* = Wo/Hy$, ordinate: $m = En/Hy$

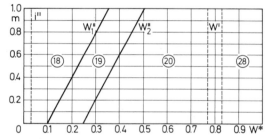

Fig. 53. $w^* - m$ graph for amphiboles in undersaturated rocks ($\Delta Q < 0$).
Abscissa: $w^* = Wo/Hy$, ordinate: $m = En/Hy$

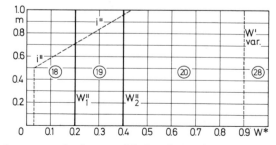

b) $\Delta Q < 0$

enter w^* and m in Fig. 53 and read the number of the next step. In both nomograms the values of w_1'' and w_2'' and i'' are given by the abscissa of the points of intersection of the ordinate m with the borderlines of the fields. Entering ΔQ in Fig. 54 one reads the values of f'', q'' and in Fig. 55, those of f', w_1' (if $k \leqq 0.4$) or w_2' (if $k > 0.4$).

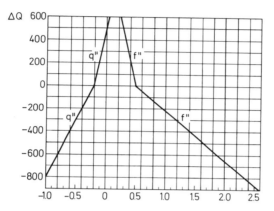

Fig. 54. Graph for the determination of the f'' and q'' factors for amphiboles

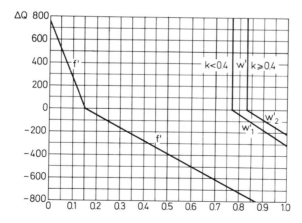

Fig. 55. Graph for the determination of the f' and w' factors for clinopyroxenes and amphiboles. $k = Or/(Or + Ab)$

(18)   Calculate: $H = 5$ Wo
and $D = f'' \cdot H/(Or + Ab + An)$.
If $D > 1$, then $-q'' \cdot H$ must be divided by D.

Il	Or	Ab	An	Wo	Hy	ΔQ	
$-i'' \cdot H$	$-D \cdot Or$	$-D \cdot Ab$	$-D \cdot An$	$-Wo$	$-H$	$-q'' \cdot H$	*Amphibole*[7]
Il*	Or*	Ab*	An*	—	Hy*	ΔQ*	⟶ (21)

(19)   Calculate: $D = f'' \cdot Hy/(Or + Ab + An)$

Il	Or	Ab	An	Wo	Hy	ΔQ	
$-i'' \cdot Hy$	$-D \cdot Or$	$-D \cdot Ab$	$-D \cdot An$	$-Wo$	$-Hy$	$-q'' \cdot Hy$	*Amphibole*[7]
Il*	Or*	Ab*	An*	—	—	ΔQ*	

   a) ΔQ* > 0, then ΔQ* = *Quartz* ⟶ (39)
   b) ΔQ* < 0, then ΔQ* = ΔQ_1 ⟶ (36)

(20)   Calculate: $H = (w' - w^*) \cdot Hy^*/(w' - w_2'')$ and $D = f'' \cdot H/(Or + Ab + An)$.
If $D > 1$, then $-q'' \cdot H$ must be divided by D.

Il	Or	Ab	An	Wo	Hy	ΔQ	
$-i'' \cdot H$	$-D \cdot Or$	$-D \cdot Ab$	$-D \cdot An$	$-w_2'' \cdot H$	$-H$	$-q'' \cdot H$	*Amphibole*[7]
Il*	Or*	Ab*	An*	Wo*	Hy*	ΔQ*	⟶ (27)

(21)   Calculate: $X = 1.2(Or^* - 0.1\ Ab^*)$ and distinguish two alternatives:
   a) $X \geq Hy^*$, then Hy''' (in biotite) = Hy* ⟶ (22)
   b) $X < Hy^*$, then Hy''' (in biotite) = X ⟶ (23)

(22)   Calculate: $k' = 0.3\,k + 0.70$

Il*	Or*	Ab*		An*	Wo = 0	Hy*	ΔQ*
-0.08 (max.)	$-0.9\,k'$	$-0.9(1 - k')$		-0.15	$+0.06$ $-Wo^{**}$	**$-Hy^*$**	$+0.52$ *Biotite* *adds Amphibole*
Il**	Or**	Ab**		An**			ΔQ**

   a) ΔQ** > 0, then ΔQ** = *Quartz* ⟶ (39)
   b) ΔQ** < 0, then ΔQ** = ΔQ_1 ⟶ (36)

---

7 The $D \cdot$ An-bearing amphiboles may be named as follows:
Hornblende, if $f'' \leq 0.75$.
Hastingsite, if $f'' > 0.75$ and $i'' \leq 0.1$.
Kaersutite, if $f'' > 1.00$ and $i'' > 0.1$.

(23)   Calculate: $k' = 0.3\,k + 0.70$

Il*	Or*	Ab*	An*	Wo*=0	Hy*	ΔQ*
−0.08 (max.)	$-0.9\,k'$	$-0.9(1-k')$	−0.15	$+0.06$ $-\text{Wo}^{**}$	−X	0.52 *Biotite* *adds Amphibole*
Il**	Or**	Ab**	An**		Hy**	ΔQ** ⟶ (24)

(24)   Distinguish three alternatives:

a) $\Delta Q^{**} > 0$, then $\Delta Q^{**} = Quartz$ and $\text{Hy}^{**} = Hypersthene$ ⟶ (39)

b) $-0.25\,\text{Hy}^{**} < \Delta Q^{**} < 0$ ————————————⟶ (25)

c) $\Delta Q^{**} < -0.25\,\text{Hy}^{**}$ ————————————————⟶ (26)

(25)

Hy**		ΔQ**
$-4|\Delta Q^{**}|$ $-\text{Hy}^{***}$	$+|\Delta Q^{**}|$	*Olivine* $= 3|\Delta Q^{**}|$ *Hypersthene* $= \text{Hy}^{***}$
—	—	————————————⟶ (39)

(26)

Hy**		ΔQ**
$-\text{Hy}^{**}$	$+0.25\,\text{Hy}^{**}$ ΔQ^{***}	*Olivine* $= 0.075\,\text{Hy}^{**}$ $= \Delta Q_1$
—	—	————————————⟶ (36)

(27)   Calculate: $k = \text{Or}/(\text{Or} + \text{Ab})$ and distinguish two alternatives. Then determine the values of x, y and z:

a) $\Delta Q \geqq 0$ b) $\Delta Q < 0$

$\left.\begin{array}{l} x = 0.3\ \text{Or} \\ y = 1.0\ \text{Ab} \\ z = 2.0\ \text{An} \end{array}\right\}$ sum = S $\left.\begin{array}{l} x = 0.2\ \text{Or} \\ y = 0.6\ \text{Ab} \\ z = 2.0\ \text{An} \end{array}\right\}$ sum = S.

In both cases proceed as follows: Enter ΔQ into Fig. 55 and read on the abscissa the value of f'.

Calculate $D = f' \cdot \text{Wo}/S$ and $\text{Or}' = x \cdot D$, $\text{Ab}' = y \cdot D$ and $\text{An}' = z \cdot D$ noting, however, that the upper limits are $\text{Or}' = \text{Or}^*$; $\text{Ab}' = \text{Ab}^*$ and $\text{An}' = \text{An}^*$.

The silica set free will be: $\text{Q}' = 0.35(\text{Or}' + \text{Ab}' + \text{An}')$. Furthermore calculate Wo^*/An^* and enter this value in Fig. 56 to read the value of p according to the alternative $k < 0.4$ or $k > 0.4$. Enter these values in the scheme:

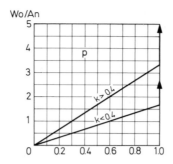

Fig. 56. Graph for the determination of the p factor

Il*	Or*	Ab*	An*	Wo*	Hy*	ΔQ*	
$-p \cdot$Il*	$-$Or'	$-$Ab'	$-$An'	$-$Wo*	$-$Hy*	$+$Q'	*Clinopyroxene*
Il**	Or**	Ab**	An**	—	—	ΔQ**	

a) $\Delta Q^{**} \geqq 0$, then $\Delta Q^{**} = Quartz$ ⟶ ③⑨

b) $\Delta Q^{**} < 0$, then $\Delta Q^{**} = \Delta Q_1$ ⟶ ③⑥

㉘ The calculations and the figures to be used are the same as in step 27, starting, however, directly from the saturated norm.

Il	Or	Ab	An	Wo	Hy	ΔQ	
$-p \cdot$Il	$-$Or'	$-$Ab'	$-$An'	$-$Wo	$-$Hy	$+$Q'	*Clinopyroxene*
Il*	Or*	Ab*	An*	—	—	ΔQ*	

a) $\Delta Q^{*} \geqq 0$, then $\Delta Q^{*} = Quartz$ ⟶ ③⑨

b) $\Delta Q^{*} < 0$, then $\Delta Q^{*} = \Delta Q_1$ ⟶ ③⑥

㉙ Calculate: $a = Ac/(Ac + Hy)$ and $w = Wo/(Ac + Hy)$ and distinguish two alternatives:

a) $\Delta Q \geqq 0$ ⟶ ③⓪

b) $\Delta Q < 0$ ⟶ ③①

㉚ Enter a and w in Fig. 57 and read the number of the next step. If the projecting point P_{aw} falls within field ㉞, then determine on the abscissa the values of w_2'' and w' corresponding to the intersection of the ordinate a with the boundaries of field ㉞.

Enter a in Fig. 58 and read on the abscissa the values of f'' and q''.

If $a > 0.4$, then the amphibole will be a *Riebeckite*,

if $a < 0.4$, then the amphibole will be an *Arfvedsonite*.

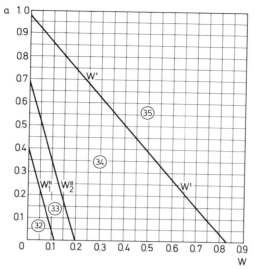

Fig. 57. $w - a$ graph for alkali amphiboles and aegirine-augites in oversaturated rocks ($\Delta Q > 0$). Abscissa: $w = \text{Wo}/(\text{Ac} + \text{Hy})$, ordinate: $a = \text{Ac}/(\text{Ac} + \text{Hy})$

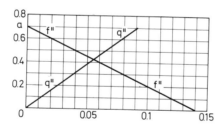

Fig. 58. Graph for the determination of the f'' and q'' factors for alkali amphiboles in oversaturated rocks ($\Delta Q > 0$). Ordinate: $a = \text{Ac}/(\text{Ac} + \text{Hy})$

③① Enter a and w in Fig. 59 and read the number of the next step. If P_{aw} falls within field ㉞, then read on the abscissa the values of w_2'' and w' corresponding to the intersections of the ordinate a with the boundary lines of field ㉞. If P_{aw} falls within field ㉜, then read the value of w_1''.

Enter a in Fig. 60 and read on the abscissa the values of f'' and q'' corresponding to the intersections of the ordinate a with the lines f'' and q''.

If w'' resp. $w_2'' < 0.15$, then the amphibole will be an *Arfvedsonite*, if $w'' > 0.15$, then the amphibole will be a *Catophorite*.

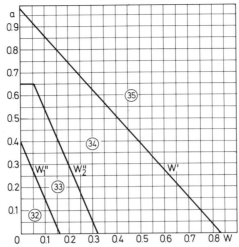

Fig. 59. $w - a$ graph for alkali amphiboles and aegirine-augites in undersaturated rocks ($\Delta Q < 0$). Abscissa: $w = Wo/(Ac + Hy)$, ordinate: $a = Ac/(Ac + Hy)$

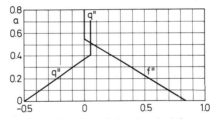

Fig. 60. Graph for the determination of the f'' and q'' factors for alkali amphiboles in undersaturated rocks ($\Delta Q < 0$). Ordinate: $a = Ac/(Ac + Hy)$

㉜   Calculate: Hy″ (in amphibole) = $5 Wo \cdot Ac$; $k' = 0.3 k + 0.70$
             $X = Ac + Hy''$ and $D = f'' \cdot X/(Or + Ab)$ [limit D = 1]

Il	Or	Ab	Ac	Wo	Hy	ΔQ	
$-0.02 \cdot X$	$-D \cdot Or$	$-D \cdot Ab$	$-Ac$	$-Wo$	$-Hy''$	$-q'' \cdot X$	*Amphibole*[8]
-0.03	$-k'$	$-(1-k')$			$-Hy^*$	$+0.52$	*Biotite*
Il*	Or*	Ab*	—	—	—	ΔQ*	

a) $\Delta Q^* \geqq 0$, then $\Delta Q^* = Quartz$ —————————————→ ㊴

b) $\Delta Q^* < 0$, then $\Delta Q^* = \Delta Q_1$ —————————————→ ㊱

8 If Ns is present, it may (at least partly) enter riebeckite or arfvedsonite. The maximum content of Ns in these amphiboles is about 20% of the calculated amount without Ns.

(33)   Calculate: $X = Ac + Hy$ and $D = f'' \cdot X/(Or + Ab)$

Il	Or	Ab	Ac	Wo	Hy	ΔQ	
$-0.02 \cdot X$	$-D \cdot Or$	$-D \cdot Ab$	$-Ac$	$-Wo$	$-Hy$	$-q'' \cdot X$	*Amphibole*[9]
Il*	Or*	Ab*	—	—	—	ΔQ^*	

a) $\Delta Q^* \geqq 0$, then $\Delta Q^* = Quartz$ ──────────────────→ (39)

b) $\Delta Q^* < 0$, then $\Delta Q^* = \Delta Q_1$ ──────────────────→ (36)

(34)   Calculate: $X = (w' - w)/(w' - w_2'')$
       $Y = X \cdot (Ac + Hy)$
       $Z = (Ac^* + Wo^* + Hy^*)$ (see the following scheme)
       $D = f'' \cdot Y/(Or^* + Ab^*)$

Il	Or	Ab	Ac	Wo	Hy	ΔQ	
$-0.02 \cdot Y$	$-D \cdot Or$	$-D \cdot Ab$	$-X \cdot Ac$	$-w_2'' \cdot Y$	$-X \cdot Hy$	$-q'' \cdot Y$	*Amphibole*[9]
$-0.015 \cdot Z$	$-0.015 \cdot Z$	$-0.125 \cdot Z$	$-Ac^*$	$-Wo^*$	$-Hy^*$	$+0.05 \cdot Z$	*Pyroxene*
Il*	Or*	Ab*	—	—	—	ΔQ^*	

If $Wo^* > Hy^*$, then calculate *sphene* according to the following scheme, in which $X = 0.5(Wo^* - Hy^*)$:

Il*	Wo*	Hy*	ΔQ^*	
$-X$	$-X$	$+X$	$-0.5 X$	*Sphene* = 1.5 X
Il**	Wo**	Hy**	ΔQ^{**}	

Distinguish two alternatives:

a) $\Delta Q^*(*) \geqq 0$, then $\Delta Q^*(*) = Quartz$ ──────────────→ (39)

b) $\Delta Q^*(*) < 0$, then $\Delta Q^*(*) = \Delta Q_1$ ──────────────→ (36)

(35)   Calculate: $X = Ac + Wo + Hy$

Il	Or	Ab	Ac	Wo	Hy	ΔQ	
$-0.015 \cdot X$	$-0.015 \cdot X$	$-0.125 \cdot X$	$-Ac$	$-Wo$	$-Hy$	$+0.05 \cdot X$	*Clinopyroxene*
Il*	Or*	Ab*	—	—	—	ΔQ^*	

If $Wo > Hy$, then calculate *sphene* according to the following scheme, in which $X = 0.5 (Wo - Hy)$:

Il*	Wo	Hy	ΔQ^*	
$-X$	$-X$	$+X$	$-0.5 X$	*Sphene* = 1.5 X
Il**	Wo**	Hy**	ΔQ^{**}	

9 See footnote on p. 118.

Distinguish two alternatives:

a) $\Delta Q^*(*) \geqq 0$, then $\Delta Q^*(*) = Quartz$ ⟶ (39)

b) $\Delta Q^*(*) < 0$, then $\Delta Q^*(*) = \Delta Q_1$ ⟶ (36)

Note: If Hl, Ah or Th are present, then calculate minerals of the sodalite group according to Key 2, step 26.

(36)   Calculate: $k^* = Or^*/(Or^* + Ab^*)$ and $q^* = |\Delta Q_1|/(Or^* + Ab^*)$.

Enter k^* and q^* in Fig. 61 and read the number of the next step. If the projecting point falls within field 37, the value of U has to be determined by interpolation along the diagonal lines. Note that the lower limit of $U = 1.50$.

(37)   Calculate: $X = (1 + U) \cdot |\Delta Q_1|$ and distinguish two alternatives:

a) $k^* \leqq 0.45$

Or*	Ab*	ΔQ_1	
$-0.33\,k^* \cdot X$	$-(1 - 0.33\,k^*) \cdot X$	$+|\Delta Q_1|$	*Nepheline*
Or**	Ab**	—	

⟶ (39)

b) $k^* > 0.45$

Or*	Ab*	ΔQ_1	
$-0.15 \cdot X$	$-0.85 \cdot X$	$+|\Delta Q_1|$	*Nepheline*
Or**	Ab**	—	

⟶ (39)

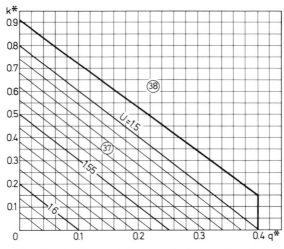

Fig. 61. $q^* - k^*$ graph for the calculation of nepheline.
Abscissa: $q^* = |\Delta Q_1|/(Or^* + Ab^*)$, ordinate: $k^* = Or^*/(Or^* + Ab^*)$

(38)   Calculate: X = 1.2 (Ab − 0.1 Or)

Or*	Ab*	ΔQ₁	
−0.15·X	−0.85·X	+0.4·X	*Nepheline*
Or**	Ab**	ΔQ₂	⟶ (39)

(39)   Calculate the average feldspar Or$_x$ Ab$_y$ An$_z$ on the basis of Or*, Ab* and An* (resp. Or**, Ab** and An**) according to the rule given in step 38 of the Key 2.

Enter x (Or) and z (An) in Fig. 62 in order to determine the nature of the feldspar or of two coexisting feldspars, *orthoclase* and *plagioclase*, and their relative amounts.

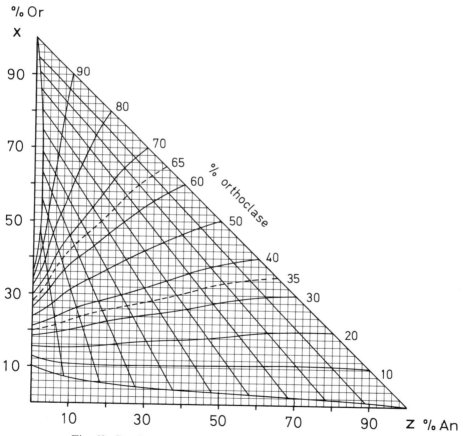

Fig. 62. Coexistence of feldspars in the pure plutonic facies

ΔQ_2 cannot be compensated by further desilication of silicate minerals and should be replaced by CO_2. However, as carbonatites, generally, do not contain amphiboles, it is better to calculate the mineral assemblage of the volcanic facies, entering then the Key for carbonatites.
Enter into Key 6 (p. 131).

Key 4: Calculation of the Mineral Assemblage of the "Dry" Subvolcanic-Plutonic Facies

① Distinguish three alternatives[10]:

a) the saturated norm contains *Sil* ⟶ ②

b) the saturated norm contains *An* and *Wo* ⟶ ⑨

c) the saturated norm contains *Ac* ⟶ ⑮

② Calculate: $h = \mathrm{Hy}/(\mathrm{Sil} + \mathrm{Hy})$ and $q = |\Delta Q|/(\mathrm{Sil} + \mathrm{Hy})$ and distinguish two alternatives:

a) Fs < 3 En; enter h and q into Fig. 63 to read the number of the next step.
b) Fs ≥ 3 En; enter h and q into Fig. 64 to read the number of the next step.

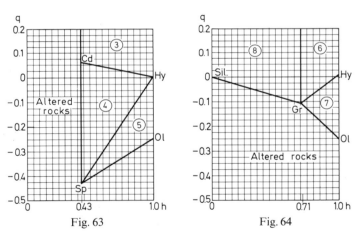

Fig. 63. $h-q$ graph with cordierite. Abscissa: $h = \mathrm{Hy}/(\mathrm{Hy} + \mathrm{Sil})$, ordinate: $q = \mathrm{Q}/(\mathrm{Hy} + \mathrm{Sil})$. Symbols see Table 21

Fig. 64. $h-q$ graph with garnet. Abscissa: $h = \mathrm{Hy}/(\mathrm{Hy} + \mathrm{Sil})$, ordinate: $q = \mathrm{Q}/(\mathrm{Hy} + \mathrm{Sil})$. Symbols see Table 21

10 Remark: if (Wo + Hy + Ac) > 2(Or + Ab + An), then use the Key for ultramafic rocks on p. 183.

(3)

Or	Ab	Sil	Hy	ΔQ	
-0.02	-0.10	$-$ **Sil**	-0.75 $-$ **Hy***	-0.10 $-$ **ΔQ***	*Cordierite* $=$ sum *Hypersthene* $=$ Hy* *Quartz* $=$ ΔQ*
Or*	Ab*	—	—	—	$\longrightarrow$ (19)

(4) Calculate: $X = 1.23\ \Delta Q + 0.88\ Sil$

Or	Ab	Sil	Hy	ΔQ	
-0.02	-0.10	$-$ **X** $-$ **Sil***	-0.75 -0.75 $-$ **Hy***	-0.10 $+0.71$	*Cordierite* $=$ sum *Spinel* $=$ sum *Hypersthene* $=$ Hy*
Or*	Ab*	—	—	—	$\longrightarrow$ (19)

(5)

Sil	Hy	ΔQ	
$-$ **Sil**	-0.75 $-4\lvert\Delta Q^{*}\rvert$ $-$ **Hy***	$+0.71$ $+$ **ΔQ***	*Spinel* $=$ sum *Olivine* $= 3\lvert\Delta Q^{*}\rvert$ *Hypersthene* $=$ Hy*
—	—	—	$\longrightarrow$ (19)

(6)

Sil	Hy	ΔQ	
$-$ **Sil**	-2.00 $-$ **Hy***	$+0.33$ $-$ **ΔQ***	*Garnet* $=$ sum *Hypersthene* $=$ Hy* *Quartz* $=$ ΔQ*
—	—	—	$\longrightarrow$ (19)

(7)

Sil	Hy	ΔQ	
$-$ **Sil**	-2.00 $-4\lvert\Delta Q^{*}\rvert$ $-$ **Hy***	$+0.33$ $+$ **ΔQ***	*Garnet* $=$ sum *Olivine* $= 3\lvert\Delta Q^{*}\rvert$ *Hypersthene* $=$ Hy*
—	—	—	$\longrightarrow$ (19)

(8)

Sil	Hy	ΔQ	
-0.5 $-$ **Sil***	$-$ **Hy**	$+0.17$ $-$ **ΔQ***	*Garnet* $=$ sum *Sillimanite* $=$ Sil* *Quartz* $=$ ΔQ*
—	—	—	$\longrightarrow$ (19)

⑨   Determine the number of the following step according to the scheme:

		$(Or+Ab) \geq 2(Wo+Hy)$	$(Or+Ab) < 2(Wo+Hy)$
Wo > Hy	$\Delta Q > -100$	step ⑩	step ⑩
	$\Delta Q \leq -100$	step ⑪	step ⑫
Wo ≤ Hy	$\Delta Q \gtrless 0$	step ⑬	step ⑬

⑩   Calculate: $X = 0.5 \, (Wo - Hy)$ and distinguish two alternatives:

a) $Il > X$

Il	Wo	Hy	ΔQ	
$-X$	$-X$	$+X$	$-0.5\,X$	$Sphene = 1.5\,X$
Il^*	$Wo^* =$	Hy^*	ΔQ^*	⟶ ⑬

b) $Il \leq X$

Il	Wo	Hy	ΔQ	
$-Il$	$-Il$	$+Il$	$-0.5\,Il$	$Sphene = 1.5\,Il$
$-$	$Wo^* =$	Hy^*	ΔQ^*	⟶ ⑬

⑪   Calculate: $X = 0.2 \, (Wo - Hy)$ and distinguish two alternatives:

a) $Il > X$

Il	An	Wo	Hy	ΔQ	
$-X$	$-X$	$-8X$	$-3X$	$+2X$	$Melanite = 11\,X$
Il^*	An^*	$Wo^* =$	Hy^*	ΔQ^*	⟶ ⑬

b) $Il \leq X$

Il	An	Wo	Hy	ΔQ^*	
$-Il$	$-Il$	$-8\,Il$	$-3\,Il$	$+2\,Il$	$Melanite = 11\,Il$
$-$	An^*	$Wo^* =$	Hy^*	ΔQ^*	⟶ ⑬

⑫   Calculate: $X = 0.5 \, (Wo - Hy)$ and distinguish two alternatives:

a) $Il > X$

Il	Wo	Hy	
$-X$	$-X$	$+X$	$Perovskite = X$
Il^*	Wo^*	Hy^*	⟶ ⑬

b) $Il \leq X$

Il	Wo	Hy	
$-Il$	$-Il$	$+Il$	$Perovskite = Il$
—	Wo*	Hy*	$\longrightarrow$ (13)

(13) Enter ΔQ into Fig. 43 and read F' on the abscissa:
If $\Delta Q > 0$, then $F' = 0.2 (1 - \Delta Q \cdot 10^{-3})$,
if $\Delta Q < 0$, then $F' = 0.2 - \Delta Q \cdot 10^3$.

Calculate $F' \cdot Wo = $ maximum amount of feldspar components (Fsp') entering clinopyroxene (Cpx). Determine the relative amounts of Or', Ab' and An' according to the proportionality schemes:

$\Delta Q \geq 0$	$\Delta Q < 0$
0.3 Or $= x$	0.2 Or $= x$
1.0 Ab $= y$	0.6 Ab $= y$
2.0 An $= z$	2.0 An $= z$
$x + y + z = S$	$x + y + z = S$

In both cases proceed as follows:
Calculate $D = F' \cdot Wo/S$, $Or' = x \cdot D$, $Ab' = y \cdot D$ and $An' = z \cdot D$ noting, however, that the upper limits are $Or' = Or$, $Ab' = Ab$ and $An' = An$.
The silica set free will be: $Q' = 0.35 (Or' + Ab' + An')$.
Calculate $k = Or/(Or + Ab)$ and the value of Wo/An.
Enter into Fig. 56 and read the value of p on the abscissa according to the alternative $k < 0.4$ or $k > 0.4$.
Then $Il' = p \cdot Il$ (upper limit $p = 1$).

Enter ΔQ into Fig. 44 and read the value of n on the abscissa with regard of k. $Hy' = n \cdot Wo$ or, if $Hy < n \cdot Wo$, then $Hy' = Hy$. In the latter case, no Hy* will be left over.

Enter the components of clinopyroxene in the following scheme:

Il	Or	Ab	An	Wo	Hy	ΔQ	
$-Il'$	$-Or'$	$-Ab'$	$-An'$	$-Wo$	$-Hy'$	$+Q'$	provisional Cpx = sum
Il*	Or*	Ab*	An*	—	Hy*	ΔQ^*	$\longrightarrow$ (14)

(14) Distinguish three alternatives according to the relative values of ΔQ^* (resp. ΔQ^{**}). Note that Hy* (resp. Hy) may be nil.

a) $\Delta Q^* > 0$:
   $\Delta Q^* = Quartz$
   $Hy^* = Hypersthene$ $\longrightarrow$ (19)

b) $-0.25 \, Hy^* \leq \Delta Q^* < 0$:
   $3|\Delta Q^*| = Olivine$
   $Hy^* - 4|\Delta Q^*| = Hypersthene$ $\longrightarrow$ (19)

c) $\Delta Q^* < -0.25\ Hy^*$:

$0.75\ Hy^* = Olivine$

$\Delta Q^* + 0.25\ Hy^* = \Delta Q_1$ —————————————————————→ ⑯

After the formation of olivine, the definite composition of clinopyroxene is known and permits the determination of the type according to the following calculation:

Calculate $c = (Wo' + 0.4\ An')/Hy'$ and split Hy' into Fs and En ($En = 2\ Mg$).

c'	$En > Fs$	$En < Fs$
> 0.9	diopside	hedenbergite
0.5 to 0.9	augite	ferroaugite
0.3 to 0.5	subcalcic augite	subcalcic ferroaugite
0.1 to 0.3	pigeonite	ferropigeonite.

Augites and diopsidic augites ($c' > 0.5$) containing more than 4 % Il in their norm (atoms) are called *titanaugites*.

Note that the ratio En/Fs increases with increasing oxidation. The above subdivision of calculated clinopyroxenes refers to the degree of oxidation resulting from the calculation of standard magnetite (Mt_0).

⑮ Let be $M = (Ac + Wo + Hy, resp.\ Hy^*)$ and calculate:

$Il' = 0.015\ M$ (upper limit: $Il' = Il$)
$Or' = 0.015\ M$; $Ab' = 0.125\ M$; $Q' = 0.050\ M$.

Enter these values into the following scheme:

Il	Or	Ab	Ac	Wo	Hy	ΔQ	
$- Il'$	$- Or'$	$- Ab'$	$- Ac$	$- Wo$	$- Hy$	$+ Q'$	*Clinopyroxene* = sum
Il^*	Or^*	Ab^*	—	—	—	ΔQ^* ——→ ⑯	

Calculate $a = Ac/M$ and distinguish the following types of alkaline clinopyroxenes:

$a \geqq 0.70 = Aegirine$
$0.70 > a > 0.15 = Aegirine\text{-}augite$
$0.15 \geqq a > 0 = Sodian\ augite$

⑯ Calculate: $k^* = Or^*/(Or^* + Ab^*)$
$\qquad\qquad q^* = |\Delta Q_1|/(Or^* + Ab^*)$

Enter k^* and q^* into Fig. 65 to read the number of the next step and, if this is No. 17, read the value of U.

⑰ Calculate: $X = (1 + U)\cdot|\Delta Q_1|$

Or*	Ab*	ΔQ_1	
$-0.33\ k^* \cdot X$	$-(1 - 0.33\ k^*) \cdot X$	$+ |\Delta Q_1|$	*Nepheline* = $1.5|\Delta Q_1|$
Or^{**}	Ab^{**}	—	——————→ ⑲

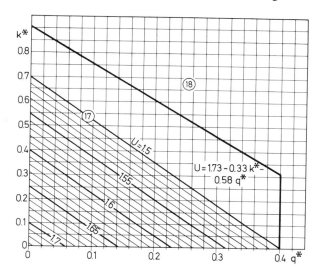

Fig. 65. $q^* - k^*$ for the calculation of nepheline. Abscissa: $q = |\Delta Q_1|/(Or^* + Ab^*)$, ordinate: $k = Or^*/(Or^* + Ab^*)$. Note: Nepheline may include up to 8% An-compound

(18) Calculate: $X = 1.2 (Ab^* - 0.1 \, Or^*)$

Or*	Ab*	ΔQ_1	
$-0.30\,X$	$-0.70\,X$	$+0.40\,X$	*Nepheline* $= 0.6\,X$
Or**	Ab**	ΔQ_2	$\longrightarrow$ (19)

Note: ΔQ_2 could be compensated by forming some *kalsilite* according to the relation: $5\,Or - 2\,Q = 3\,Ks$, or by forming carbonates (see Key 5). Which one of these compensations should be used is indicated by the mode.

(19) Calculate the "average feldspar" $Or_x \, Ab_y \, An_z$ on the basis of the feldspar compounds Or* (resp. Or.**), Ab* (resp. Ab**) and An* left over after the formation of clinopyroxene, biotite, and feldspathoids.
Denoting $Or^* + Ab^* + An^* = Fsp^*$ (resp. $Or^{**} + Ab^{**} + An^* = Fsp^*$) determine:

$$x = 100 \, Or^* / Fsp^*$$
$$y = 100 \, Ab^* / Fsp^*$$
$$z = 100 \, An^* / Fsp^* \longrightarrow (20)$$

Enter x and z into Fig. 49 to determine the nature of the feldspars and their amount. Then enter into Key 6 (p. 131).

Key 5: Approximate Calculation of Carbonatites[11]

(1) Distinguish two alternatives:

   a) the saturated norm contains An and Wo ⟶ (2)

   b) the saturated norm contains Ac ⟶ (7)

(2) Distinguish two alternatives:

   a) Wo > Hy ⟶ (3)

   b) Wo ≦ Hy ⟶ (6)

(3) Distinguish two alternatives:

   a) (Or + Ab) ≧ (Wo + Hy) ⟶ (4)

   b) (Or + Ab) < (Wo + Hy) ⟶ (5)

(4) Calculate: $X = 0.2$ (Wo − Hy) and distinguish two alternatives:

   a) Il > X:

Il	An	Wo	Hy	ΔQ	
−X	−X	−8X	−3X	+2X	*Melanite* = 11 X
Il*	An*	Wo*	Hy*	ΔQ*	⟶ (6)

   b) Il ≦ X:

Il	An	Wo	Hy	ΔQ	
−Il	−Il	−8 Il	−3 Il	+2 Il	*Melanite* = 11 Il
—	An*	Wo*	Hy*	ΔQ*	⟶ (6)

(5) Calculate: $X = 0.5$ (Wo − Hy) and distinguish two alternatives:

   a) Il > X:

Il	Wo	Hy	
−X	−X	+X	*Perovskite* = X
Il*	Wo*	= Hy*	⟶ (6)

   b) Il ≦ X:

Il	Wo	Hy	
−Il	−Il	+Il	*Perovskite* = Il
—	Wo*	= Hy*	⟶ (6)

---

11 Only to be used, if in Key 2 melilite has been formed

(6) Calculate: $x = 0.2$ Or, $y = 0.6$ Ab and $z = 2$ An. Enter ΔQ into Fig. 43 and read the value of F'.
Calculate: $D = F' \cdot Wo/(x + y + z)$ and determine: $Or' = D \cdot x$; $Ab' = D \cdot y$, $An' = D \cdot z$ and $Q' = 0.35 (Or' + Ab' + An')$. The upper limits are: $Or' = Or$, $Ab' = Ab$ and $An' = An$.
Calculate: $w = Wo/Hy$ and determine: $Il' = w \cdot Il$ (upper limit = Il).

Il	Or	Ab	An	Wo	Hy	ΔQ		
$-$ Il'	$-$ Or'	$-$ Ab'	$-$ An'	$-$ Wo	$-$ Wo	$+Q'$	provisional Cpx$_0$ = sum	
Il*	Or*	Ab*	An*	—	Hy*	ΔQ^*	⟶	(8)

(7) Calculate: $X = (Ac + Wo + Hy)$ and determine:
$Or' = 0.015 X$; $Ab' = 0.125 X$; $Il' = 0.015 X$ and $Q' = 0.05 X$

Il	Or	Ab	Ac	Wo	Hy	ΔQ		
$-$ Il'	$-$ Or'	$-$ Ab'	$-$ Ac	$-$ Wo	$-$ Hy	$+Q'$	prov. Cpx$_0$ = sum	
Il*	Or*	Ab*	—	—	—	ΔQ_1	⟶	(10)

(8) Distinguish: a) Hy* has been left over ⟶ (9)
b) Hy* = 0 ⟶ (10)

(9) Calculate: $k = Or/(Or + Ab)$ and $k' = 0.3 k + 0.70$ and distinguish:
a) $Or^* \geqq 1.2$ Hy*

Il*	Or*	Ab*	Hy*	ΔQ^*		
-0.08	$-0.9 k'$	$-0.9 (1-k')$	$-$ **Hy***	$+0.52$	*Biotite* = sum	
Il**	Or**	Ab**	—	ΔQ_1	⟶	(10)

b) $Or^* < 1.2$ Hy*

Il*	Or*	Ab*	Hy*	ΔQ^*		
-0.07	$-$**0.8 Or***	-0.17	-1.18	$+0.61$	*Biotite* = sum	
			$-$ **Hy****	$+0.25$	*Olivine* = 0.75 Hy**	
Il**	Or**	Ab**	—	ΔQ_1	⟶	(10)

Note: Distribute Il proportionally between clinopyroxene and biotite.

(10) Calculate: $X = 1.2 (Ab^* - 0.1 Or^*)$

Or*	Ab*	An*	ΔQ_1		
$-0.15 X$	$-0.85 X$		$+0.40 X$	*Nepheline* = 0.6 X	
Or**	Ab**	An*	ΔQ_c	⟶	(11)

⑪ On the basis of the remaining feldspar compounds, calculate the average feldspar $Or_x Ab_y An_z$.

Enter x and z into Fig. 62 and determine the nature and the amounts of feldspars (orthoclase $\pm$ plagioclase).

Note that ΔQ_C remains unchanged. ───────────────→ ⑫

⑫ Convert the clinopyroxene completely into provisional silicates and carbonates, distinguishing two alternatives:

a) the clinopyroxene contains An' ───────────────→ ⑬

b) the clinopyroxene contains Ac ───────────────→ ⑭

⑬ Provisional clinopyroxene $= Il' + Or' + Ab' + An + Wo + Hy' - Q'$
(as before)

Provisional ilmenite Il_0 $= Il^*$
Provisional orthoclase Ort_0 $= 1.2\ Or'$
Provisional nepheline Ne_0 $= 0.6\ (Ab' - 0.2\ Or')$
Provisional spinel Sp_0 $= 0.6\ An'$
Provisional carbonates $Carb_0$ $= Wo + Hy'$

$\Sigma Q_C = 0.5\ carb_0 + 0.667\ (Ne_0 + Sp_0) - Q'$ ───────────────→ ⑮

⑭ Provisional clinopyroxene $Cpx_0 = Il' + Or' + Ab' + Ac' + Wo + Hy' - Q'$
Provisional ilmenite Il_0 $= Il^*$
Provisional orthoclase Ort_0 $= 1.2\ Or'$
Provisional nepheline Ne_0 $= 0.6\ (Ab' - 0.2\ Or')$
Provisional carbonates $Carb_0$ $= Wo + Hy'$

$\Sigma Q_C = 0.5\ Carb_0 + 0.667\ Ne_0 - Q'$ ───────────────→ ⑮

⑮ Calculate: $R = |\Delta Q_C|/\Sigma Q_C$ and the final minerals:

Orthoclase $=$ orthoclase $+ R \cdot Ort_0$
Plagioclase $=$ plagioclase (often lacking)
Nepheline $=$ nepheline $+ R \cdot Ne_0$
Augite $= (1 - R) \cdot Cpx_0 + R \cdot Ac$
Biotite $=$ biotite
Melanite $=$ melanite (mostly lacking)
Perovskite $=$ perovskite (often lacking)
Ulvöspinel etc. $=$ magnetite $+ R \cdot (Il_0 + Sp_0)$
Apatite $=$ apatite
Carbonates $=$ Calcite $+ R \cdot Carb_0$.

Note: Besides these orthomagmatic minerals, most carbonatites contain secondary minerals of pneumatolytic or hydrothermal origin, as e.g. cancrinite, sulfide ores, minerals of rare elements (dysanalite, etc.) or zeolites, secondary carbonates etc.

Enter into Key 6 (p. 131).

Key 6: Determination of Volume Percent and of the Name of the Rock

In order to express the mineral assemblage as volume percent, multiply the figures calculated as numbers of atoms for each mineral by the following factors (in alphabetic order):

amphiboles	1.03	magnetite	0.91
anorthoclase	1.25	marialite	1.19
apatite	1.14	melanite	0.99
biotite	1.10	melilite	1.12
breunnerite	1.23	muscovite	1.20
calcite	1.11	nepheline	1.08
cancrinite	1.17	nosean	1.28
clinopyroxenes	**1.00**	olivine	0.88
cordierite	1.22	perovskite	1.02
corundum	0.77	plagioclases	1.21
cossyrite	0.96	pyrite	1.42
fayalite	0.88	pyroxenes	**1.00**
garnet	0.86	quartz	1.36
haüyne	1.28	sanidine, orth.	1.28
hercynite	0.91	scapolites	1.24
hypersthene	**1.00**	sillimanite	1.00
ilmenite	0.93	sodalite	1.28
kalsilite	1.17	sphene (titanite)	1.12
kyanite	0.90	spinel	0.91
leucite	1.33	zircon	1.22

The sum of the volume equivalents, thus obtained, is reduced to one hundred. The resulting volume percent of the single minerals are given to one decimal place.

The result of these calculations represent the *stable mineral assemblages* of the various facies which approach closely the modal composition of the rock. The modes of mixed facies can be interpolated (see p. 84).

Determination of the Rock Name

In order to establish the field of the double-triangle in which the rock in question falls, denote:

Q_0 = vol.-% of quartz.

A_0 = vol.-% of sanidine + anorthoclase or orthoclase.

P_0 = vol.-% of plagioclase.

F_0 = vol.-% of nepheline + leucite + sodalite group.

CI = colour index = sum of femic minerals and accessories.

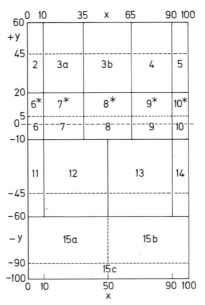

Fig. 66. Graph of the determination of the field of the Streckeisen system in which the rock in question is falling

If CI ≥ 90, then the rock will fall into field 16,
if CI < 90, then calculate the following coordinates:
$$x = 100 \ P_0/(A_0 + P_0) \ ,$$
$$+y = 100 \ Q_0/(Q_0 + A_0 + P_0) \ ,$$
$$-y = 100 \ F_0/(F_0 + A_0 + P_0) \ .$$
Enter x and $\pm y$ in the graph of Fig. 66 and read the number of the field in which the projected point P_{xy} is falling.

The names of volcanic rocks according to the classification proposed by STRECKEISEN are to be read in Table 19. Those of plutonic rocks

Table 19. Classification of *volcanic rocks* (modified after A. STRECKEISEN, 1967 and 1972)

CI = Colour Index. If CI is smaller than indicated, then use the prefix *leuco*, if it is greater, then use the prefix *mela*.

Field No.	Name of the family	CI	Varieties and remarks
2	Alkali (feldspar) rhyolite	0–10	Soda rhyolite: if the rock contains soda-rich minerals (Na-sanidine, anorthoclase, Na-pyroxenes, etc.)

Table 19 (continued)

Field No.	Name of the family	CI	Varieties and remarks
3a	Rhyolite	0–15	
3b	Rhyodacite	0–20	
4	Dacite	5–25	
5	Plagidacite	10–30	Instead of the old name quartz-andesite
6*	Alkali (feldspar) quartz-trachyte	0–20	Soda quartz-trachyte: if the rock contains soda-rich minerals (cf. soda rhyolite)
6	Alkali (feldspar) trachyte	0–20	Soda trachyte: if the rock contains soda-rich minerals (cf. soda rhyolite)
6	Foid-bearing alkali (feldspar) trachyte	0–20	Foid-bearing soda trachyte: (cf. soda rhyolite)
7*	Quartz-trachyte	5–25	
7	Trachyte	5–25	
7	Foid-bearing trachyte	5–25	
8*	Quartz-latite	10–35	
8	Latite	10–35	
8	Foid-bearing latite	10–35	

A. If $\sigma^a < 4$ and $\tau^b > 9$

Field No.	Name of the family	CI	Varieties and remarks
9*	Quartz-latiandesite	15–38	Mela quartz-latiandesite: if CI > 38 (instead of quartz-latibasalt).
9	Latiandesite	15–38	Mela latiandesite: if CI > 38 (instead of latibasalt).
9	Foid-bearing latiandesite	15–38	Rare.

B. If $\sigma > 4$ and $\tau \leqq 9$ or if $\sigma < 4$ and $\tau < 9$

Field No.	Name of the family	CI	Varieties and remarks
9*	Quartz-mugearite	15–38	Rare.
9	Mugearite	15–38	Quartz-bearing mugearite: if Q < 5. Olivine mugearite: if Ol > 5. As most of the mugearites contain foids the adjective foid-bearing is unnecessary.
9*	Quartz-latibasalt	38–70	
9	Latibasalt	38–70	

[a] $\sigma = (Na_2O + K_2O)^2/(SiO_2 - 43)$.

[b] $\tau = (Al_2O_3 - Na_2O)/TiO_2$

In both cases weight percentages in the analysis.

Table 19 (continued)

Field No.	Name of the family	CI	Varieties and remarks

A. If $\sigma < 4$ and $\tau > 9$

10*	Quartz-andesite	20–38	Mela quartz-andesite: if CI > 38 (instead of quartz-basalt).
10	Andesite	20–38	Olivine andesite: if Ol > 5. Mela andesite: if CI > 38 (instead of basalt).
10	Foid-bearing andesite	20–38	Rare.

B. If $\sigma > 4$ and $\tau \lesseqqgtr 9$ or if $\sigma < 4$ and $\tau < 9$

10*	Quartz-hawaiite	20–38	Rare.
10	Hawaiite	20–38	Quartz-bearing hawaiite: if $Q < 5$. Olivine hawaiite: if Ol > 5. As most of the hawaiites contain foids the adjective foid-bearing is unnecessary.
10*	Tholeiitic quartz-basalt	38–70	
10	Tholeiitic basalt	38–70	Mostly quartz-bearing ($Q < 5$) or free of quartz or foids. Olivine-bearing tholeiitic basalt: if Ol < 10 but $F = 0$.
10	Tholeiitic olivine basalt	38–70	If Ol > 10 but $F = 0$.
10	Alkali basalt	38–70	Contains always foids ($F < 10$). Olivine-bearing alkali basalt: if Ol < 10.
10	Alkali olivine basalt	38–70	If Ol > 10.
11	Phonolite	0–25	$F = 10 - 45$: According to the content in nepheline, leucite,
12	Tephriphonolite	10–40	sodalite, haüyne, nosean, etc., use in the case of the fields 11–14
13	Phonotephrite	20–50	the corresponding names, e.g. Nepheline phonolite, Sodalite-
14	Tephrite	30–70	bearing nepheline tephriphonolite (if Sod < 5), Nepheline leucite phonotephrite (if Ne < Lc), Leucite tephrite, etc. $F = 45 - 60$: Nepheline-rich phonolite, Nepheline-rich sodalite phonote- phrite, etc. Olivine tephrite ("basanite"): if Ol > 10. Carbonate-bearing tephrite: if *primary* carbonates < 5. Olivine carbonate tephrite: if *primary* carbonates > 5.

Table 19 (continued)

Field No.	Name of the family	CI	Varieties and remarks
15a	Phononephelinite Phonoleucitite	5–45	Olivine phononephelinite, Olivine tephrileucitite, etc.: if Ol > 10.
15b	Tephrinephelinite Tephrileucitite	30–60	If sodalite, haüyne, etc. are present, distinguish: Sodalite-bearing phononephelinite (Sod < 5), Carbonate tephrinephelinite (primary carbonates > 5), etc.
15c	Nephelinite Leucitite	40–70	Haüyne-bearing nephelinite, Sodalite olivine nephelinite, Melilite leucitite (Mel > 5), etc.
16	Picrite (Ol > Py + Mel)	90–100	E.g.: Melilite picrite, Nepheline melilitite,
	Ankaratrite (Py > Ol + Mel)	90–100	Kalsilite melilitite, Olivine melilitite, etc.
	Melilitite (Mel > Ol + Py)	90–100	
	Carbonatite	90–100	True carbonatites are subvolcanic rocks except the alkali carbonatite of Oldoinyo Lengai volcano/ E-Africa.

Note: The names of varieties are given as examples. In many instances it is commendable to indicate the nature of the plagioclase or that of the clinopyroxene, e.g. andesine dacite, labradorite dacite, or pigeonite dacite, aegirine-augite phonolite, etc.

according to the classification proposed by STRECKEISEN are given in Table 20.

Stress must be laid on the fact (discussed on p. 70) that heteromorphic volcanic and plutonic rocks may not fall into the same field. In order to denominate the magma type, the name of the volcanic rock has to be used and not that of the corresponding plutonic rock.

Table 20. Classification of *plutonic rocks* (modified after A. STRECKEISEN, 1967 and 1972)

CI = Colour Index. If CI is smaller than indicated, then use the prefix *leuco*, if it is greater, then use the prefix *mela*.

Field No.	Name of the family	CI	Varieties and remarks
1	Silexite	0–10	
2	Alkali (feldspar) granite	0–20	Biotite alkali (feldspar) granite (if Bi > 5). Riebeckite-bearing alkali (feldspar) granite (if Rieb < 5).
3a, b	Granite	5–20	Field 3a: syenogranite, field 3b: monzogranite. Hornblende granite (if Hbl > 5).
4	Granodiorite	5–25	If An < 50. Biotite hornblende granodiorite (if Bi < Hbl).
4	Granogabbro	20–40	If An > 50.
5	Trondjemite	0–15	
	Tonalite	15–40	
6*	Alkali (feldspar) quartz-syenite	0–25	
6	Alkali (feldspar) syenite	0–25	Lusitanite: if CI = 50–75.
6	Foid-bearing alkali (feldspar) syenite	0–25	According to the content in nepheline, sodalite, etc. use in the case of the fields 6–10 the corresponding names, e.g. nepheline-bearing syenite, sodalite- and nepheline-bearing monzonite, etc.
7*	Quartz-syenite	10–35	
7	Syenite	10–35	
7	Foid-bearing syenite	10–35	
8*	Quartz-monzonite	15–45	Hornblende biotite quartz-monzonite (Hbl < Bi).
8	Monzonite	15–45	Biotite monzonite (Bi > 5,
8	Foid-bearing monzonite	15–45	else biotite-bearing monzonite, etc.).
9*	Quartz-monzodiorite	20–50	If An < 50.
9*	Quartz-monzogabbro	25–60	If An > 50.
9	Monzodiorite	20–50	Olivine-bearing monzodiorite: if Ol < 5.
9	Monzogabbro	25–60	Olivine monzogabbro: if Ol > 5.

Table 20. (continued)

Field No.	Name of the family	CI	Varieties and remarks
9	Foid-bearing monzodiorite	20–50	If An < 50.
9	Foid-bearing monzogabbro	25–60	If An > 50.
10*	Quartz-diorite	25–50	If An < 50.
10*	Quartz-gabbro	35–65	If An > 50.
10	Anorthosite	0–10	A detailed subdivision of the plutonic rocks falling into the field 10 is given in Fig. 67a–c.
10	Diorite	25–50	If An < 50.
10	Gabbro	35–65	If An > 50. Olivine gabbro, etc.: see Fig. 67a–c.
10	Foid-bearing diorite	25–50	If An < 50.
10	Foid-bearing gabbro	35–65	If An > 50.
11	Foid alkali (feldspar) syenite	0–30	Instead of the name foyaite. $F = 10$–45 (Field 11–14): Nepheline alkali (feldspar) syenite,
11	Malignite	30–60	Sodalite-bearing nepheline alkali (feldspar) syenite (if Sod < 5), etc.
11	Shonkinite	60–90	$F = 45$–60 (Field 11–14): Nepheline-rich alkali (feldspar) syenite, etc.
12	Foid monzosyenite	15–45	Instead of the name plagifoyaite. Sodalite nepheline monzosyenite (Sod < Ne), etc.
13	Essexite	30–60	Nepheline essexite ($F < 45$) Nepheline-rich essexite, etc. ($F = 45$–60).
14	Theralite	35–65	Carbonate-bearing theralite (if *primary* carbonates < 5).
15a	Foyafoidite	10–40	Foyanephelinite, etc.
15b	Therafoidite	35–65	Carbonate theranephelinite (if *primary* carbonate > 10).
15c	Foidite	0–90	Nephelinite, Sodalite nephelinite, etc.
	Urtite	0–30 ⎫	
	Ijolite	30–60 ⎬	If Na ≫ K
	Melteigite	60–90 ⎭	
	Italite	0–30 ⎫	
	Fergusite	30–60 ⎬	If K ≫ Na
	Missourite	60–90 ⎭	
16	Peridotites Pyroxenites Hornblendites	90–100	A detailed subdivision of the plutonic rocks falling into the field 16 is given in Fig. 68a, b.
16	Carbonatite	90–100	

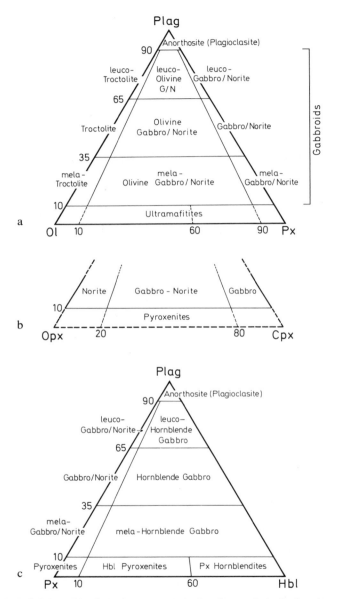

Fig. 67. Subdivision of basic rocks, composed of mafites and plagioclase (An > 50 %), falling into field 10 of the Streckeisen double-triangle. a With olivine, both pyroxenes and plagioclase. b Subdivision of gabbro and norite. Subdivision into leuco or "normal" or mela rocks of this group according to Fig. 67a. c Basic rocks with hornblende

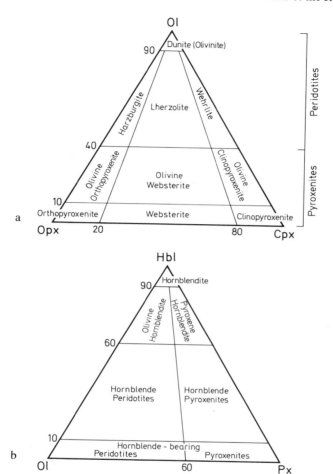

Fig. 68. Subdivision of ultramafic rocks (ultramafitites), falling into field 16 of the Streckeisen double-triangle. a Composed essentially of olivine, orthopyroxene and clinopyroxene. With accessory garnet: if garnet <10%: Garnet-bearing ultramafitites, if garnet >10%: Garnet ultramafitites. b With hornblende. Subdivision according to the special systematic of peridotites and pyroxenites in Fig. 68a, e.g.: Orthopyroxene hornblendite (60–90 Hbl, 10–40 Opx), hornblende lherzolite (40–90 Ol + Px, 10–60 Hbl), hornblende-bearing olivinite (>90 Ol, <10 Hbl)

IX. Examples

In the following, some selected examples are calculated in detail and commented on in order to illustrate not only the procedure of calculating but also the critical interpretation of the results and the calculation of heteromorphic mineral assemblages in different facies.

It must be emphasized that the present calculations should not be used uncritically. Their principal aim is to furnish a basis for studying the relations between chemical and mineralogical composition of the rocks. The results of the calculations yield a basis, a kind of coordinates, to which the observable facts can be referred to.

For the sake of clearness, the examples are calculated by slide rule neglecting the decimals. Using a computer or any type of calculator yielding decimals, the results may be slightly different. However, these differences will hardly ever exceed the unavoidable analytical errors.

Example No. 1: "Hawaiian Tholeiitic Basalt"

(Average of 181 analyses; G. A. MACDONALD and T. KATSURA 1964, p. 124, no. 8)

Key 1

Weight %		Atoms	Ap	Il	Mt_0	Or	Ab	An	Wo	Hy	ΔQ
SiO_2	49.36	822				24	207	197	79	310	+5
Al_2O_3	13.94	274				8	69	197			
Fe_2O_3	3.03	38									
FeO	8.53	119		31	27					101	
MnO	0.16	2									
MgO	8.44	209								209	
CaO	10.33	184	6					99	79		
Na_2O	2.13	69					69				
K_2O	0.38	8				8					
TiO_2	2.50	31		31							
P_2O_5	0.26	4	4								
		1760	10	62	27	40	345	493	158	620	+5

Calculation of the Saturated Norm

step 1: Multiply the weight percent of the analysis (1st column) by the respective factors to obtain the amounts of atoms (column 2). The decimals are ommitted.

step 2: Degree of oxidation: $Ox^\circ = 38/159 = 0.24$.

step 3: $\sigma = 2.51^2/6.36 = 0.99$ $\tau = (13.94 - 2.13)/2.50 = 4.73$.

step 7: $Il > 25$. From Fig. 40 results: $Mt_0 = 0.212 (159-31) = 27$.

Key 2

Saturated Norm

Ap	Il	Mt_0	Or	Ab	An	Wo	Hy	ΔQ	
10	62	27	40	345	493	158	620	+ 5	sum = 1780 (control)
	−11				−8	−24	−158	−620	+11	*clinopyroxene* = 810
	51		40	337	469	—	—	+16	(remainder) quartz = 16
			−40	−337	−469				*feldspar* = 846
			—	—	—				

Distribution of the components of the saturated norm among the minerals to be calculated (Volcanic Facies):

step 1: Alternative b) (with An and Wo).

step 14: $Wo = 158 < Hy = 620$ leads to step 22.

step 22: From Fig. 43 one obtains $F' = 0.20$,   hence $F'Wo = 32$.
Since $\Delta Q > 0$, calculate x, y, z, and reduce them proportionally to the sum of $Fsp' = F'Wo$ according to the following scheme:

$x = 0.3$	Or $=$ 12	$0 = Or'$	rounded up to
$y = 1.0$	Ab $=$ 345	$8 = Ab'$	integer numbers
$z = 2.0$	An $=$ 986	$24 = An'$	
	sum $= 1343$	$32 = Fsp'$	

$Q' = 0.35\ Fsp' = 11$

Calculate $k = 40/385 = 0.104$ and $Wo/An = 158/493 = 0.32$.
From Fig. 56 it results that $p = 0.17$, hence $Il' = 0.17 \times 67 = 11$.
Enter ΔQ into Fig. 44, from which results $n = 1.3$, being $k < 0.4$.
Note that, in this case, n is not needed in the volcanic facies, but it will be used in further discussions.

step 24: Alternative a),

step 26: Alternative a),

step 27: Alternative a):   *Quartz* = 18, *Hypersthene* = $Hy^* = 0$.

step 38: Calculate the average feldspar according to the following proportionality scheme:

$Or^* =$ 40	$x =$ 5
$Ab^* = 337$	$y =$ 40
$An^* = 469$	$z =$ 55
sum $= 846$	$\% = 100$

step 39: Entering x and z into Fig. 49 it is seen that only one feldspar is formed, namely a *labradorite* with An = 55 %.

Key 6

Atoms		Factor	Vol. equ.	Vol.-%		
Quartz	16	1.36	20	1.0	$Q = 1.9$	$y = 1.9$
Labradorite	842	1.21	1024	52.9	$P = 98.1$	$x = 100.0$
Pigeonite	810	1.00	810	41.8		
Magnetite	27	0.91	25	1.3	CI = 46.1	
Ilmenite	51	0.92	47	2.4		
Apatite	10	1.14	11	0.6		
	1760		1937	100.0		

Multiply the number of atoms of each mineral (column 1) by the respective factor (column 2) to obtain the volume equivalents (column 3) which, reduced to the sum 100, yield the *volume percents of the stable mineral assemblage* (column 4) as the final result.

Calculate the coordinates of the rock in the double-triangle, reducing the salic minerals (in the present example quartz and plagioclase) to the sum $100 = P + Q$. For convenience calculate also the parameters $x = 100$ and $y = 1.9$ to be entered in Fig. 66. The projecting point P_{xy} falls into field 10 on the P-Q-border. Since the colour index CI = 46.3 and the value $\tau = 4.73$, the rock is a typical *tholeiitic basalt*.

Control of the Calculation

The sum of atoms (1760) equals the sum of the components of the saturated norm and the sum of atoms attributed to the different minerals.

Comment

Tholeiitic basalts often contain phenocrysts of hypersthene and augite, formed under subvolcanic to plutonic conditions, where pigeonite, being unstable under higher pressure, cannot form.

The maximum amount of the two pyroxenes can be calculated as follows:

Hypersthene $= Hy - n \cdot Wo = 620 - 13 \times 158 = 415$ atoms.
Augite $=$ pigeonite $-$ hypersthene $= 810 - 415 = 395$ atoms, or
Hypersthene (maximum) $= 21.4$ Vol.-%,
Augite (minimum) $= 20.4$ Vol.-%.

These maximum amounts are obtained also by calculating the mineral assemblage of the "dry" subvolcanic-plutonic facies with the aid of *Key 4*. The resulting rock would be a *two pyroxene gabbro*.

In the volcanic rock, only a part of the pyroxene will appear as phenocrysts of hypersthene and augite, whereas the remainder will form microlites of pigeonite or may be present only virtually in a glassy matrix.

Example No. 2: "Alkalic Basalt"

(Hawaiian group, average of 29 analyses; G. A. MACDONALD and T. KATSURA 1964, p. 124, no. 4)

Weight %		Atoms	Ap	Il	Mt_0	Or	Ab	An	Wo	Hy	ΔQ	
SiO_2	46.46	774				54	282	175	88	301	-126	
Al_2O_3	14.64	287				18	94	175				
Fe_2O_3	3.27	41										
FeO	9.11	127		38	34					98		
MnO	0.14	2										
MgO	8.19	203								203		
CaO	10.33	184	8					88	88			
Na_2O	2.92	94					94					
K_2O	0.84	18				18						
TiO_2	3.01	38		38								
P_2O_5	0.37	5	5									
		1773	13	76	34	90	470	438	176	602	-126	
				-19		-1	-14	-43	-176	-206	$+22$	$Augite = 437$
	remainder:		57			89	456	395	—	396	-104	
										-396	$+99$	$Olivine = 297$
	remainder:					89	456	395		-5		
						-1	-12			$+5$		$Nepheline = 8$
	remainder = Fsp*					88	444	395				
						$Or_{9\frac{1}{2}}$	Ab_{48}	$An_{42\frac{1}{2}}$				

Sanidine $= 2\% = 18$ *Andesine* $= 927 - 18 = 909$

Atoms		Factor	Vol. equ.	Vol.-%		
Sanidine	18	1.28	23	1.2	$A = 2.0$	
Andesine	909	1.21	1100	56.9	$P = 97.2$	$x = 98.0$
Nepheline	8	1.08	9	0.5	$F = 0.8$	$y = -0.8$
Augite	437	1.00	437	22.7		
Olivine	297	0.88	261	13.6		
Magnetite	34	0.91	31	1.6	$CI = 41.4$	
Ilmenite	57	0.92	52	2.7		
Apatite	13	1.14	15	0.8		
	1773		1928	100.0		

Result: *Alkali basalt*

Notes

The calculation has been summarized in one scheme, i.e. without re-writing the saturated norm. The various steps of the calculation are the following:

Key 1

step 1: Analogous to the 1st example.
step 2: $Ox^0 = 41/170 = 0.241$.
step 3: $\sigma = 3.76^2/3.46 = 4.08$.
$\tau = (14.64-2.92)/3.01 = 3.90$.
step 5: Alternative b) .
step 7: $Il > 25$, hence $Mt_0 = 1.2 \,(0.1 + 0.112) \cdot (170 - 38) = 34$.

Key 2

step 1: Alternative b) .
step 14: Wo < Hy, leads to step 22.
step 22: From Fig. 43: $F' = 0.33$, $F'Wo = 58$

$0.2\ Or = x =$ 18	$1 = Or'$	
$0.6\ Ab = y =$ 282	$14 = Ab'$	
$2.0\ An = z =$ 876	$43 = An'$	
1176	58	$Q' = 0.35 \times 58 = 20$

$k = 90/560 = 0.161$, $Wo/An = 176/438 = 0.40$,
from Fig. 56: $p = 0.25$, hence $Il' = 0.25 \times 76 = 19$,
from Fig. 44: $n = 1.17$, hence $Hy' = 1.17 \times 176 = 206$.
step 24: Alternative a).
step 26: Alternative a).
step 27: Alternative c), remainder $\Delta Q_1 = -5$.
step 28: $k^* = 89/546 = 0.163$ and $q^* = 5/546 = 0.009$,
from Fig. 46: $U = 1.653$.
step 29: $X = 2.653 \times 5 = 13$ and $0.33 \times 0.164 \times 13 =$ about 1 (Or″)
and $13 - 1 = 12$ (Ab″), $U \cdot |\Delta Q_1| = 8$.
step 38:

Or* = 88	$9^1/_2 = Or'$	
Ab* = 444	$48\quad = Ab'$	average feldspar
An* = 395	$42^1/_2 = An''$	
927	100	

From Fig. 49 it results that *sanidine* = 2%, hence *andesine* = 98% of the remaining *feldspar* (922).

Key 6

see example Nr. 1.

Comment

The term "alkalic basalt" refers to the nature of the plagioclase which – in contrast to the labradorite in tholeiitic basalts – is an andesine. Another characteristic is the high content in olivine.

The "dry" subvolcanic-plutonic facies shows the same mineral assemblage as that of the volcanic facies. Hypersthene cannot form because all Hy* will be desilicated to form olivine. For the same reason the clinopyroxene appears to be rich in Ca and in Tschermak molecules. Its content in Fe depends, naturally, on the degree of oxidation in the early stage of crystallization. The higher the oxygen fugacity the less iron will enter the pyroxene. If the magnetite is standardized (Mt_0) as in the present calculation, then $c = (176 + 17)/206 = 0.95$. On the other hand a considerable amount of ilmenite (Il') enters the clinopyroxene: $Il' = 19/437 = 4.35\%$. These values characterize the clinopyroxene as a diopsidic augite just sufficiently titaniferous to be called titanaugite.

Example No. 3: "High-alumina Basalt"

(Average of 11 analyses; H. KUNO 1960, p. 141)

Weight %		Atoms	Ap	Il	Mt_0	Or	Ab	An	Wo	Hy	ΔQ	
SiO_2	50.19	836				24	267	248	60	274	−37	
Al_2O_3	17.58	345				8	89	248				
Fe_2O_3	2.48	36										
FeO	7.19	100		9	39					91		
MnO	0.25	3										
MgO	7.39	183								183		
CaO	10.50	187	3					124	60			
Na_2O	2.75	89					89					
K_2O	0.40	8				8						
TiO_2	0.75	9		9								
P_2O_5	0.14	2	2									
		1798	5	18	39	40	445	620	120	548	−37	
			−2				−5	−23	−120	−152	+11	$Cpx(1) = 291$
	remainder:			16		40	440	597	—	396	−26	
										+104	+26	$Olivine = 78$
	remainder:					40	440	597		292		$Cpx(2) = 292$
	= Fsp* = 1077					Or_4	Ab_{41}	An_{55}				$Labradorite = 1077$

Atoms		Factor	Vol. equ.	Vol.-%	
Labradorite	1077	1.21	1303	64.8	$P = 100$
Pigeonite	583	1.00	583	29.0	
Olivine	78	0.88	69	3.4	
Magnetite	39	0.91	35	1.7	$CI = 35.2$
Ilmenite	16	0.92	15	0.8	
Apatite	5	1.14	6	0.3	
	1798		2011	100.0	

Result: *Olivine-bearing andesite*

Notes

The calculation is similar to that of the previous example and does not need a detailed explanation. Only a few remarks will be sufficient.

Key 1

step   7: $Il < 25$, hence $Mt_0 = 0.3 \, (133 - 9) = 39$
(Fig. 40).

Key 2

step 27: Alternative b) . $Hy^{**} = 396 - 104 = 292$ adds Cpx.
The final clinopyroxene consists hence of two parts:
$Cpx \, (1) = 291 + Cpx \, (2) = 292 = 583$,
$c = (120 + 9)/(152 + 292) = 0.290$, i.e. pigeonite .

Key 1

step   2: $Ox^0 = 0.28$.
step   3: $\sigma = 1.38$ and $\tau = 19.8$.

Comment

According to the preferred nomenclature, this volcanic rock is an *andesite* ($CI < 38$). According to ROSENBUSCH's systematics it would be a basalt, because the plagioclase is labradorite. However, the high value of τ and the low value of σ show clearly the great affinity to andesite and the irrelation to true basalt (see p. 9).

Among the phenocrysts two pyroxenes (augite and hypersthene) may occur in about equal amounts.

Example No. 4: "Two Pyroxene Andesite"

(Iō-jima volcano, Japan; H. MATSUMOTO [in K. ONO 1962, p. 417, no. 1037])

Weight %		Atoms	Ap	Il	Mt₀	Or	Ab	An	Wo	Hy	ΔQ	
SiO₂	67.78	1129				111	360	110	13	73	462	
Al₂O₃	13.62	267				37	120	110				
Fe₂O₃	1.99	25										
FeO	3.60	50		4	29					43		
MnO	0.08	1										
MgO	1.19	30								30		
CaO	4.50	80	12					55	13			
Na₂O	3.71	120					120					
K₂O	1.74	37				37						
TiO₂	0.29	4		4								
P₂O₅	0.49	7	7									
		1750	*19*	8	*29*	185	600	275	26	146	462	
			—			−2	−1	−26	−34	1		*Cpx* (1) = 62
Fsp* = 1057				8		185	598	274	—	112	*463*	
San. % = 18						Or₁₇½	Ab₅₆½	An₂₆		−112		*Cpx* (2) = 112

Atoms		Factor	Vol. equ.	Vol.-%		
Quartz	463	1.36	630	29.3	Q = 32.8	y = +32.8
Sanidine	190	1.28	243	11.3	A = 12.7	
Andesine	867	1.21	1049	48.7	P = 54.5	x = 81.1
Pigeonite	174	1.00	174	8.1		
Ti-magnetite	37	0.91	34	1.6	CI = 10.7	
Apatite	19	1.14	22	1.0		
	1750		2152	100.0		

Result: *Pigeonite dacite*

Notes

Il < 25:   Mt₀ = 0.408 (76 − 4) = 29 (see Fig. 40).

The calculation of clinopyroxene (Key 2, step 22) has already been explained in the first example. As ΔQ > 0, the determination of $n = 1.3$ is not needed for the volcanic facies, because all Hy will enter the clinopyroxene. In the scheme, Hy* is calculated separately, because it indicates the amount of hypersthene which would form under "dry" plutonic conditions, and which, partly, may appear as intratelluric phenocrysts in the volcanic rock, presenting a mixed facies. $Hy^* = 112$ is the maximum amount of hypersthene, corresponding to 5.2% volume, besides augite (2.9%).

Comment

The name "two pyroxene andesite", given by MATSUMOTO to the lava of Iō-jima, is a typical pheno-name, because the rock is in reality a dacite. Wishing to express both, the mode and the true character of the rock, one should call it a *"pheno-andesitic two pyroxene dacite"*.

The heteromorphic plutonic rock can be calculated by the aid of Key 3 as follows:

Ap	Il	Mt_0	Or	Ab	An	Wo	Hy	ΔQ	
19	8	*29*	185	600	275	26	146	462	(saturated norm, sum = 1750)
	−3					−3	+3	−1	*Sphene* = 4
	5		185	600	275	23	149	461	
	−3		−6	−21	−10	−23	−123	−1	*Hornblende* = 187 + 2 = 189
	2		179	579	265	—	26	460	
	−2		−18	−5	−4	+2	−26	+13	*Biotite* = 40
	—		161	574	261	(2)	—	473	*Quartz* = 473
$Fsp^* =$			Or_{16}	Ab_{58}	An_{26}	=		996	*Orthoclase* (20%) = 199
									Andesine (80%) = 797

Atoms		Factor	Vol. equ.	Vol.-%		
Quartz	473	1.36	643	29.9	$Q = 34.6$	$y = +34.6$
Orthoclase	199	1.28	255	11.8	$A = 13.6$	
Andesine	797	1.21	965	44.8	$P = 51.8$	$x = \;\;79.2$
Hornblende	189	1.03	195	9.1		
Biotite	40	1.10	44	2.0		
Magnetite	29	0.91	26	1.2	$CI = 13.5$	
Sphene	4	1.12	4	0.2	field 4	
Apatite	19	1.14	22	1.0		
	1750		2154	100.0		

Result: *Biotite-bearing hornblende granodiorite*

Notes

step 12: $w = 26/146 = 0.178$, $X = 0.356 \times 8 = 2.8$.

step 17: $w^* = 23/149 = 0.154$, $m = 60/146 = 0.41$,   alternative a.

step 18: $H = (0.214 \times 149)/0.26 = 123$, $D = (0.3 \times 123)/1060 = 0.0348$.

step 21: $X = 1.2 (179{-}58) = 145$, $Hy^* = 26$, $w_1'' = 0.20$,

   $f'' = 0.30$, $q'' = -0.01$.

step 22: $k' = 0.071 + 0.700 = 0.771$, $0.9\, Hy^* = 23.4$.

Comment

The heteromorphic relations between dacite and hornblende grano-diorite show only slight differences as far as concerns the systematic position of the two mineral facies. The plutonic rock shows a higher colour index, because hornblende and biotite contain more salic compounds than pigeonite. As biotite contains "Or", one could expect a lower percentage of orthoclase in the granodiorite, but the contrary results from the calculation, because the gap of miscibility among the feldspars has increased.

The plutonic mineral assemblage demonstrates that the same magma can yield pigeonite dacites, two pyroxene dacites, hornblende dacites or biotite hornblende dacites, according to prevailing conditions during the occasional formation of intratelluric phenocrysts. Only *pigeonite dacite* represents the pure volcanic facies.

Two pyroxene dacite is a mixture between volcanic and subvolcanic facies, whereas *hornblende dacite* or *biotite hornblende dacite* represent mixed facies between volcanic and plutonic assemblages. Under "dry" plutonic or subvolcanic conditions *two pyroxene granodiorite* will be formed which differs from two pyroxene dacite only in structural features, whereas under "wet" plutonic conditions *biotite-bearing hornblende granodiorite* will originate.

The calculation of mixed facies will be illustrated by example no. 8 ("Hornblende basalt" from Wickersberg).

The same procedure of calculating could be applied also for the rocks of the present example.

Example No. 5: "Cordierite-bearing Rhyolite"

(Ignimbrite; Roccastrada/Tuscany/Italy; R. MAZZUOLI (1967, Tab. 5, no. 136)

Weight %		Atoms	Ap	Il	Mt_0	Or	Ab	An	Sil	Hy	ΔQ	
SiO_2	72.08	1200				315	297	28	15	24	521	
Al_2O_3	13.38	262				105	99	28	30			
Fe_2O_3	1.17	15										
FeO	0.76	11		2	7					17		
MnO	0.02	—										
MgO	0.30	7								7		
CaO	1.03	18	4					14				
Na_2O	3.08	99					99					
K_2O	4.93	105				105						
TiO_2	0.15	2		2								
P_2O_5	0.19	<3	3									
		1722	7	4	7	525	495	70	45	48	521	
				−1		−13	−2		−4	−17	+9	Biotite = 28
				3		512	493	70	41	31	530	
						−1	−4		−41	−31	−4	Cordierite = 81
						511	489	70	—	—	526	Quartz = 526
						Or_{48}	$Ab_{45\frac{1}{2}}$	$An_{6\frac{1}{2}}$				Sanidine = 995
												Oligoclase = 75

Atoms		Factors	Vol. equ.	Vol.-%		
Quartz	526	1.36	713	31.9	$Q = 34.3$	$y = +34.3$
Sanidine	995	1.28	1274	57.0	$A = 61.3$	
Oligoclase	75	1.21	91	4.1	$P = 4.4$	$x = 6.7$
Cordierite	81	1.22	107	4.8		
Biotite	28	1.10	31	1.4	$CI = 7.0$	
Ore	10	0.91	9	0.4		
Apatite	7	1.14	8	0.4		
	1722		2233			

Result: *Cordierite-bearing alkali rhyolite*

Notes
Key 1
step 5: Alternative a).
step 6: Na + K = 204. From Fig. 40 results: $X_1 = 0.304$.
Fe − Ti = 24, hence: $Mt_0 = 24 X_1 =$ about 7.

Key 2

step 1: Alternative a).

step 2: $k = 525/1020 = 0.515$,
$h = 48/93 = 0.516$,
$q = 521/93 = 5.60$.

step 11: $k' = 0.154 + 0.7 = 0.854$,
$X = 57.6 - 40.5 = 17$.

step 38: Average feldspar $= Or_{48} Ab_{45\frac{1}{2}} An_{6\frac{1}{2}}$.
From Fig. 49 results 93% sanidine; Fsp* $= 1070$.
$0.93 \times 1070 = 995$ sanidine and 75 oligoclase.

Comment

The mode has been determined as follows:

Phenocrysts:	quartz	14.5
	sanidine	15.7
	plagioclase	9.5
	cordierite	1.7
	biotite	5.2
	granophyric matrix	53.4

The calculated result corresponds qualitatively to the mode, but there are quantitative differences which have to be explained.

The calculated amounts of plagioclase and biotite are too low, whereas the amount of cordierite is too high.

This discrepancy between calculated minerals and modal phenocrysts may be due to alteration of the rock. In fact, the feldspar in the granophyric groundmass looks turbid from the presence of minute flakes of sericite, and hydrated iron ore is dispersed in the matrix and found in thin veinlets. Evidently, some alkalis and iron have been lost during the degassing of the ignimbrite sheet.

These losses can be calculated approximately as follows:

a) Saturated Norms of the Phenocrysts

Calculate the amount of atoms according to the relation:

$$\text{Atoms} = \frac{\text{Volume-\%} \times \text{sum of volume equivalents of the rock}}{100 \times \text{volume factor}},$$

i.e. in the present example:

$$\text{Quartz} = \frac{14.5 \times 2233}{136} = 239,$$

$$\text{Sanidine} = \frac{15.7 \times 2233}{128} = 275 \ (Or_{67} \ Ab_{30} \ An_3),$$

$$\text{Plagioclase} = \frac{9.5 \times 2233}{121} = 175 \ (Or_5 \ Ab_{65} \ An_{30}) \ ,$$

$$\text{Biotite} = \frac{5.2 \times 2233}{110} = 106 \ (\text{see Key 2, step 11}) \ ,$$

$$\text{Cordierite} = \frac{1.7 \times 2233}{122} = 31 \ (\text{see Key 2, step 11}) \ .$$

	Atoms	Il	Or	Ab	An	Sil	Hy	ΔQ
Quartz	239	—	—	—	—	—	—	239
Sanidine	275	—	184	83	8	—	—	—
Plagioclase	175	—	9	114	52	—	—	—
Biotite	106	4	49	8	—	14	63	−32
Cordierite	31	—	—	1	—	16	12	2
Phenocrysts	826	4	242	206	60	30	75	209

b) Saturated Norm of the Matrix

	Il	Or	Ab	An	Sil	Hy	ΔQ
Norm of the rock	4	525	495	70	45	48	521
− phenocrysts	−4	−242	−206	−60	−30	−75	−209
Norm of the matrix	—	283	289	10	15	−27	312
− 38 sericite (Ms)	—	−23	−5	—	−15	—	+5
	—	260	284	10	—	−27	317
$+ 27 \ FeO - 13^1/_2 \ Q$	—	—	—	—	—	+27	$-13^1/_2$
Quartz $= 303^1/_2$	—	260	284	10		—	$303^1/_2$
Sanidine $= 554$							

Mineral Assemblage of the Matrix

	Atoms	Factor	Vol. equ.	Vol.-%
Quartz	$303^1/_2$	1.36	412	18.9
Sanidine	554	1.28	709	32.4
Sericite	38	1.20	46	2.1
	895		1167	53.4

Conversion of sericite into sanidine according to the relation:

$$7 \text{ Ms} + 6 \text{ Q} + 2(\text{K, Na}) = 15 \text{ sanidine},$$

resp. $$38 \text{ Ms} + 33 \text{ Q} + 9 \text{ K} + 2 \text{ Na} = 82 \text{ sanidine}.$$

From these calculations it results that about

$$9 \text{ K} + 2 \text{ Na} + 13^1/_2 \text{ Fe have been lost by leaching.}$$

Mineral assemblage of the fresh rock = phenocrysts + reconstructed granophyric groundmass:

	Atoms	Factor	Vol.equ.	Vol.-%		
Quartz: 239 + 303	542	1.36	737	32.3	$Q = 34.8$	$y = 34.8$
Sanidine: 275 + 554 + 82	911	1.28	1166	51.0	$A = 55.1$	
Plagioclase:	175	1.21	212	9.3	$P = 10$	$x = 15.4$
Cordierite	31	1.22	38	1.7		
Biotite:	106	1.10	117	5.1		
Magnetite	7	0.91	6	0.3	$CI = 7.4$	
Apatite	7	1.14	8	0.3		
	1779		2284	100.0		

Result: *Cordierite-bearing biotite rhyolite*

A similar cordierite-bearing rhyolite from Roccatederighi (Tuscany) has been studied by the same author (MAZZUOLI 1967, Tab. 5, no. 107). Its mode has been determined as follows:

Phenocrysts of quartz	16.6
Phenocrysts of sanidine	19.7
Phenocrysts of plagioclase	6.9
Phenocrysts of cordierite	3.1
Phenocrysts of biotite	3.7
Granophyric matrix	50.0

The phenocrysts are fresh, but the feldspar of the granophyric groundmass is partly sericitised and hydrated iron ore is finely disseminated in the matrix and concentrated in thin veinlets. Evidently, the rock has been altered by fumarolic gases. The calculation of the mineral assemblage according to Key 2 yields:

Weight %		Atoms	Ap	Il	Mt$_0$	Or	Ab	An	Sil	Hy	ΔQ	
SiO$_2$	72.68	1210				312	153	16	56	15	658	
Al$_2$O$_3$	14.40	283				104	51	16	112			
Fe$_2$O$_3$	1.11	14										
FeO	0.49	7		8	3					10		
MgO	0.20	5								5		
CaO	0.56	10	2					8				
Na$_2$O	1.58	51					51					
K$_2$O	4.90	104				104						
TiO$_2$	0.65	8		8								
P$_2$O$_5$	0.11	2	2									
		1694	4	16	3	520	255	40	168	30	658	
						−1	−4		−40	−30	−4	Cordierite = 79
				16		519	251	40	128		−654	
						−197	−40		−128		+42	Muscovite = 323
Fsp* = 573						322	211	40			696	Sanidine = 544
Sanidine = 95 %						Or$_{56}$	Ab$_{37}$	An$_7$				Andesine = 29
												Quartz = 696

	Atoms	Factor	Vol.equ.	Vol.-%
Quartz	696	1.36	947	43.3
Sanidine	544	1.28	696	31.9
Andesine	29	1.21	35	1.6
Cordierite	79	1.22	96	4.4
Muscovite	323	1.20	388	17.8
Iron ore	19	0.92	17	0.8
Apatite	4	1.14	5	0.2
	1694		2184	100.0

The result differs considerably from the mode, because the matrix of the rock is altered, whereas the phenocrysts are fresh. An attempt can be made to calculate the modal composition of the rock by calculating first the saturated norm of the phenocrysts and subtracting the result from the saturated norm of the rock. The difference will represent the composition of the matrix.

$$\text{Quartz} = \frac{16.6 \times 2184}{1.36} = 267 \text{ (Q)},$$

$$\text{Sanidine} = \frac{19.7 \times 2184}{1.28} = 338 \text{ (Or}_{67}\text{Ab}_{30}\text{An}_3\text{)},$$

$$Plagioclase = \frac{6.9 \times 2184}{1.21} = 130 \, (Or_5Ab_{65}An_{30}) \, ,$$

$$Cordierite = \frac{3.1 \times 2184}{1.22} = 56 \, (Or_1Ab_5Sil_{51}Hy_{38}Q_5) \, ,$$

$$Biotite = \frac{3.7 \times 2184}{1.10} = 74 \, (Il_5Or_{47}Ab_7Sil_{13}Hy_{59}Q_{-31}) \, .$$

The average composition of cordierite and biotite is deduced from the scheme given in step 11 of Key 2. The following saturated norms of the phenocrysts are proportional to the above formulae but rounded off.

Phenocrysts	Atoms	Il	Or	Ab	An	Sil	Hy	ΔQ
Quartz	267							267
Sanidine	338		227	101	10			
Plagioclase	130		6	85	39			
Cordierite	56			3		29	21	3
Biotite	74	4	35	5		10	43	−23
Sum	865	4	268	194	49	39	64	247

Subtracting member by member from the saturated norm of the rock, one obtains the saturated norm of the matrix, namely:

	Il	Or	Ab	An	Sil	Hy	ΔQ
Matrix 822 =	12	252	61	−9	129	−34	411

The excess of 129 Sil and the deficiency of −34 Hy and −9 An are a consequence of the alteration of the matrix. The negative figures indicate that some cations have been lost by leaching as shown by the following compensations:

$$-34 \, Hy + 17 \, Q + 17 \, Fe = 0 \quad (17 \, Fe \text{ have been lost})$$
$$- 9 \, An + 6 \, Sil + 2 \, Q + 2 \, Ca = 1 \, An \quad (2 \, Ca \text{ have been lost})$$

The remainder must be distributed among quartz, sanidine and sericite, using the average composition of muscovite $(Or_{61}Av_{12}Sil_{40}Q_{-13})$ being deduced from step 12 of Key 2.

	Il	Or	Ab	An	Sil	ΔQ	
Remainder 829 = *12*		252	61	1	123	392	
		−188	−37		−123	+40	Sericite (Ms) = 308
		64	24	1	—	432	Quartz = 432
		Or$_{72}$	Ab$_{27}$ An$_1$				Sanidine = 89

Furthermore, there are 15 iron ore and 4 apatite to be added. According to the observed mode, the volume of the matrix equals 50%. Its mineral composition can be calculated as follows:

	Atoms	Factor	Vol.equ.	Vol.-%
Quartz	432	1.36	588	27.0
Sanidine	89	1.28	114	5.2
Sericite	308	1.20	370	17.0
Iron ore	15	0.92	14	0.6
Apatite	4	1.14	5	0.2
	848		1091	50.0

To get an idea about the original mineral composition of the unaltered rock, the sericite must be converted into sanidine according to the relation:

$$7 \, Ms + 6 \, Q + 2(K, Na) = 15 \text{ sanidine} ,$$

i.e. $$308 \, Ms + 264 \, Q + 88(K, Na) = 660 \text{ sanidine} .$$

Quartz will be reduced: $432 - 264 = 168$ corresponding to 10.5 vol.-%. Sanidine will be: $89 + 660 = 749$, corresponding to 38.7 vol.-%.

The resulting mineral assemblage of the unaltered rock is:

Vol.-%	Phenocrysts	+ Matrix	= Total rock
Quartz	16.6	10.5	27.1
Sanidine	19.7	38.7	58.4
Plagioclase	6.9	—	6.9
Cordierite	3.1	—	3.1
Biotite	3.7	—	3.7
Iron ore	—	0.6	0.6
Apatite	—	0.2	0.2
	50.0	50.0	100.0
$Q = $ 29.4	$A = 63.1$		$P = $ 7.5
$y = +29.4$			$x = 10.6$

Result: *Cordierite-bearing biotite rhyolite*

Example No. 6: "Hyalo-Pantellerite" 157

The present example illustrates an attempt to re-integrate the mineral composition of an altered rock into that of the fresh rock; it shows furthermore that the calculation of the analysis of an altered rock is senseless in itself.

Example No. 6: "Hyalo-Pantellerite"

(Pantelleria/Italy; R. Romano 1969, p. 696)

	Weight %	Atoms	Hl	Ap	Il	Or	Ab	Ac	Ns	Wo	En	Fs	ΔQ	
SiO_2	68.00	1132				300	366	78	37	8	11	29	303	
Al_2O_3	11.34	222				100	*122*							
Fe_2O_3	3.16	39						*39*						
FeO	2.42	34			8							29		
MnO	0.21	3												
MgO	0.45	11									11			
CaO	0.53	9		1						8				
Na_2O	7.76	250	14				122	39	75					
K_2O	4.72	100				*100*								
TiO_2	0.65	8			8									
P_2O_5	0.06	<1		*1*										
Cl	0.51	(14)	—											
		1809	*14*	2	16	500	610	156	112	16	22	58	303	
						−16			−24			−56	−8	*Cossyrite* = 104
						500	610	156	*88*	16	22	2	295	
						−4	−30	−156		−16	−22	−2	+12	*Aegirine* = 218
Fsp* = 1076						496	580	—		—	—	—	307	*Quartz* = 307
						Or₄₆	Ab₅₄							*Sanidine* = 1076

	Atoms	Factor	Vol.equ.	Vol.-%	
Quartz	307	1.36	418	18.9	$Q = 23.3$ $y = +23.3$
Sanidine	1076	1.28	1377	62.1	$A = 76.7$ $x = 0$
Aegirine	218	1.00	218	9.8	
Cossyrite	104	0.96	100	4.5	$CI = 14.4$
Apatite	2	1.14	2	0.1	
Sodasilite	88	1.00	88	4.0 ⎫	
Halite	14	1.00	14	0.6 ⎭	dissolved in the glass = 4.6
Control:	1809		*2217*	100.0	

Result: *Cossyrite aegirine-augite rhyolite* (peralkaline rhyolite)

Notes
Key 1

step 2: $Ox^0 = 39/76 = 0.513$.
step 3: $\tau = 12.48^2/(68 - 43) = 6.17$
$\tau = (11.34 - 7.76)/0.65 = 5.50$.
step 4: $Hl = 14$; remaining $Na = 250 - 14 = 236$.
step 5: Alternative d) .
step 9: Norm containing Ac and Ns .

Key 2

step 1: Alternative d) .
step 21: Alternative b): $Il/Fs = 16/58 = 0.277 =$ case A)
and $Ns = 112 > 24.94 = 0.43 \cdot Fs$.
step 23: $M = 252$, $a = 156/252 = 0.62$,   aegirine-augite.
step 26: Alternative b) A).
step 27: Alternative a) $Hy^* = 0$.
step 38: $Fsp^* = 1076$; $Or_{46} Ab_{54} An_0 =$ Sanidine.

Comment

Chemically the rock is a *peralkaline rhyolite*, i.e. sodasilite (Ns) partly enters cossyrite. The remaining Ns and the Hl are dissolved in the glass.

With respect to common rhyolites the σ of persodic rhyolites is high and the τ is very low, but these values are not indicative of the origin of the persodic magmas. However, it may be mentioned that the supply of Na and Ti compounds by gaseous transfer increases σ and lowers τ considerably.

Plutonic facies corresponding to "Hyalo-Pantellerite"
Calculating by Key 3: steps $1 - 29 - 30 - 34 - 39$.
Saturated norm:

Hl	Ap	Il	Or	Ab	Ac	Ns	Wo	Hy	ΔQ	
14	*2*	16	500	610	156	112	16	80	303	Sum = 1809
		−4	−1	−1	−123	−40	−2	−63	−16	*Riebeckite* = 210 + 40 Ns
		12	499	609	33	72	14	17	287	
		−1	−1	−8	−33		−14	−17	+3	*Aegirine-augite* = 71
		11	498	601	—		—	—	290	*Quartz* = 290
										Orthoclase = 1099
Remainder: Hl = 14 and Ns = 72										*Ilmenite* = 11

	Atoms	Factor	Vol.equ.	Vol.-%		
Quartz	290	1.36	394	18.4	$Q = 21.9$	$y = +21.9$
Orthoclase	1099	1.28	1407	65.7	$A = 78.1$	$x = 0$
Riebeckite	250	1.03	258	12.0		
Aegirine-augite	71	1.00	71	3.3		
Ilmenite	11	0.92	10	0.5	$CI = 15.9$	
Apatite	2	1.14	2	0.1		
	1723		2142	100.0		
Loss of Hl, Ns	86					
Control	1809					

Result: *Riebeckite granite*

Notes
Key 3

step 1: Alternative c).
step 29: Alternative a).
step 30: $a = 156/236 = 0.66$, $w = 16/236 = 0.07$,
   $w_2'' = 0.01$, $w' = 0.27$, $f'' = 0.008$, $q'' = 0.086$.
step 34: $X = 0.20/0.26 = 0.77$, $Y = 0.77 \times 236 = 182$,
   $Z = 123 + 63 + 2 = 188$ (cf. scheme),
   $D = (0.008 \times 182)/1110 = 1.46/1110 = 0.0013$.
   Note that to the sum of riebeckite (210) maximally about 20 % Ns can be added; so riebeckite $= 210 + 40 = 250$ (see footnote to step 34).

Comment

Riebeckite granite is found among the rare ejected blocks at Pantelleria. Its mode corresponds to the result of the calculation.

The loss of Hl and Ns during crystallization has been demonstrated by ROMANO (1969), by comparing the analyses of the glassy crust and of the nearly holocrystalline inner part of a pantellerite lava flow.

Example No. 7: "Sodalite-nepheline-phonolite"

(Psaccasinaptic lava dome of Mt. Campagnano, Ischia/Italy;
V. GOTTINI, unpublished)

Weight %		Atoms	Hl	Ap	Il	Or	Ab	Ac	Ns	Wo	En	Fs	ΔQ		
SiO$_2$	60.10	1001				381	681	50	9	18	16	13	−167		
Al$_2$O$_3$	18.03	354				127	227								
Fe$_2$O$_3$	2.04	25						25							
FeO	1.29	18			10							13			
MnO	0.32	5													
MgO	0.63	16									16				
CaO	1.05	19		1						18					
Na$_2$O	8.70	281	10				227	25	19						
K$_2$O	6.00	127				127									
TiO$_2$	0.78	10			10										
P$_2$O$_5$	0.06	1		1											
Cl	0.35	(10)	(10)												
(without Cl)		1854	10	2	20	635	1135	100	28	36	32	26	−167		
Cossyrite					−8				−11			−26	−4	=	49
			10		12	635	1135	100	*17*	36	32	—	−171		
Aegirine-augite					−3	−3	−24	−100		−36	−32		+9	=	189
			10		9	632	1111	—		—	—		−162		
Sodalite			−10				−150						+60	=	100
			—			632	961						−102		
Nepheline						−34	−228						+102	=	160
						598	733						—		
Sanidine						Or$_{45}$	Ab$_{55}$							=	1331

	Atoms	Factor	Vol.equ.	Vol.-%		
Sanidine	1331	1.28	1704	75.0	$A = 85.0$	$x = 0$
Nepheline	163	1.08	176	7.8	$F = 15.0$	$y = -15.0$
Sodalite	99	1.28	127	5.6		
Aegirine-augite	206	1.00	206	9.1		
Cossyrite	49	0.96	47	2.1		
Ilmenite	9	0.92	8	0.3	CI = 11.6	
Apatite	2	1.14	2	0.1	field 11	
	1857		2270	100.0		

Result: *Nepheline (peralkaline) phonolite*

Notes
Key 2

step 21: $4 (Ac + Wo + Hy) = 776 < 1770 = Or + Ab \rightarrow b)$,
$Il/Fs = 20/26 = 0.77 \rightarrow B)$,
$Ns = 28 \quad 11.4$.
step 23: $M = 168$.
step 26: $D \cdot Cl = 9.9$, sodalite $= 99$.
step 27: $Hy^* = 0$.
step 28: $k = 635/1770 = 0.355 =$ alternative a,
$k^* = 632/1598 = 0.395$,
$q^* = 104/1598 = 0.065$.
step 29: $U = 1.567$,
$X = 2.567 \times 104 = Ne + |\Delta Q_1| = 267$,
$0.33\ k^* = 0.132$.
step 38: $Fsp^* = 1331 \quad Or_{45} Ab_{55}$.

Comment

The analysed specimen represents the lava of a great exogenic dome built up by completely welded spatters. Modally it is an alkaline phenotrachyte with about 6 % phenocrysts of sanidine and about 9 % small crystals of aegirine-augite, cossyrite, sanidine and ore in a glassy matrix. In fact, it is a peralkaline phonolite.

Before the formation of the dome, pumice together with some blocks of plutonic rocks were ejected during the first phase of the eruption. Among these blocks are found foyaitic types which contain alkali feldspars, sodalite, nepheline, aegirine-augite, alkaline amphibole and opaque ore. In order to establish if they are heteromorphic equivalents of the pheno-alkali trachyte, the wet plutonic facies must be calculated.

Key 3
Plutonic Facies

Hl	Ap	Il	Or	Ab	Ac	Ns	Wo	Hy	ΔQ	
10	2	20	635	1135	100	28	36	58	−167	
	−1				−29	−11	−3	−17	−2	*Arfvedsonite* = 63
10		19	635	1135	71	17	33	41	−169	
		−2	−2	−18	−71	−17	−33	−41	+7	*Aegirine-augite* =177
10		*17*	633	1117	—	—	—	—	−162	
−10				−148					+59	*Sodalite* = 99
			633	969					−103	
			−35	−227					+103	*Nepheline* = 159
			598	742					—	
			Or$_{45}$	Ab$_{55}$						*Orthoclase* = 1340

	Atoms	Factor	Vol.equ.	Vol.-%	
Orthoclase	1340	1.28	1715	75.4	$A = 85.1$ $x = 0$
Sodalite	99	1.28	127	5.6	$F = 14.9$ $y = -14.9$
Nepheline	159	1.08	172	7.6	
Aegirine-augite	177	1.00	177	7.8	
Arfvedsonite	63	1.03	65	2.8	$CI = 11.4$
Ilmenite	17	0.92	16	0.7	
Apatite	2	1.14	2	0.1	field 11
	1857		2274	100.0	

Result: *Sodalite-bearing foyaite*

Notes
Key 3

step 1: Alternative c) .

step 29: $a = 100/158 = 0.633$,

  $w = 36/158 = 0.288$.

  Alternative b) .

step 31: $w_2'' = 0.065$, $w' = 0.295$ (arfvedsonite) ,

  $f'' = 0$, $q'' = +0.05$.

step 34: $X = (0.295 - 0.228)/(0.295 - 0.065) = 0.292$,

  $Y = 0.292 \times 158 = 46.1$,

  $Z = (Ac + Wo + Hy) = 194$ (being $f'' = 0$) ,

  $D = 0$.

step 36: $k* = 633/1602 = 0.395$,
  $q* = 103/1602 = 0.064$,
  $U = 1.545$.
step 37: $X = 2.545 \times 103 = 262$,
  $0.33\ k* = 0.132$.
step 39: The soda-rich orthoclase can easily undergo ex-solution, yielding a perthite with about 40 % albite.

Comment

The result of the calculation demonstrates that the above mentioned ejected blocks are in fact heteromorphic equivalents of the peralkaline phonolite. In some of these blocks, the alkaline feldspar is ex-solved to perthite with about 40 % albite.

Example No. 8: "Leuko-Hornblende-Olivinbasalt"
(Wickersberg/W-Germany; W. AHRENS and R. VILLWOCK
1966, p. 316)

Weight %		Atoms	Ah	Cc[12]	Ap	Il	Mt_0	Or	Ab	An	Wo	Hy	ΔQ	
SiO₂	40.59	675						57	312	159	96	366	−315	
Al₂O₃	14.39	282						19	104	159				
Fe₂O₃	5.11	64												
FeO	8.16	114				44	37					100		
MnO	0.19	3												
MgO	10.73	266										266		
CaO	10.74	192	1	3	13					79	96			
Na₂O	3.21	104							104					
K₂O	0.91	19						19						
TiO₂	3.48	44				44								
P₂O₅	0.58	8			8									
CO₂	0.15	3		3										
Cl	0.02	(−)												
So₃	0.07	1	1											
		1775	2	6	21	88	37	95	520	397	192	732	−315	
						−26		−2	−29	−73	−192	−192	+36	*Ti-augite* = 478
						62		93	491	324	—	540	−279	
			−2					−4	−26				+12	*Haüyne* = 20
								89	465	324		540	−267	
												−540	+135	*Olivine* = 405
								89	465	324	—	—	−132	= ΔQ_1
								−18	−318				+132	*Nepheline* = 204
Fsp* = 542								71	147	324			—	*Sanidine* = 43
								Or_{13}	Ab_{27}	An_{60}				*Labradorite* = 499

12 Calcite is of secondary origin and not taken into account, hence sum = 1769

	Atoms	Factor	Vol.equ.	Vol.-%			
Sanidine	43	1.28	55	3.0	$A = 6.1$		
Labradorite	499	1.21	604	32.6	$P = 66.7$	$x =$	91.6
Nepheline	204	1.08	220	11.9	$F = 27.2$	$y = -27.2$	
Haüyne	20	1.28	26	1.4			
Titanaugite	478	1.00	478	25.7			
Olivine	405	0.88	356	19.2	$CI = 51.1$		
Magnetite	37	0.91	34	1.8			
Ilmenite	62	0.92	57	3.1	field 14		
Apatite	21	1.14	24	1.3			
	1769		1854	100.0			
+ Sec. Calc.	6						
	= 1775						

Result: *Nepheline olivine tephrite (basanite)*

Key 1

step 2: $Ox^0 = 0.36$.
step 3: $\sigma = -7.04$,
$\quad \tau = 3.21$.
step 7: $Mt_0 = 0.27 (181 - 44) = 37$.

Key 2

step 22:
$x =$	19	2
$y =$	312	29
$z =$	796	73
	1127	104

$S = 615 < 1804 = 2M$,
$F'Wo = 0.54 \times 192 = 104$,
$Q' = 36$, $n = 1$,
$Wo/An = 192/397 = 0.484$, $p = 0.29$,
$Il' = 26$

step 24: $k = 95/615 = 0.154$ (v is not needed).
step 25: Alternative b) .
step 26: Alternative d). The small amount of Cl ($= 0.6$ atoms) is neglected because it enters apatite.
step 27: Alternative c) .
step 28: $k = 0.154$ (see step 22) < 0.4, hence alternative a.
$\quad k^* = 89/554 = 0.161$, $q^* = 182/554 = 0.238$,
$\quad U = 1.543$.
step 29: $X = 2.543 \times 132 = 336$, $0.33 \, k^* = 0.053$.
step 38: $Fsp^* = 542$

Or* =	71	13 = x
Ab* =	147	27 = y
An* =	324	60 = z
Fsp* =	542	100

Sanidine = 8 % = 43, Labradorite = 92 % = 499 .

Comment

The name "Leuko-Hornblende-Olivinbasalt" given by AHRENS and VILLWOCK is based on the phenocrysts and microphenocrysts: plagioclase, hornblende, olivine and augite, with *basic* plagioclase (hence basalt according to ROSENBUSCH) strongly predominating (hence erroneously "Leuko").

Modally the rock is a *pheno-hornblende olivine basalt*, but its chemical composition is that of a *nepheline basanite*.

The presence of about 18% modal hornblende indicates a mixed facies. At first, the "wet" plutonic facies must be calculated by the aid of Key 3, starting from the saturated norm and neglecting SO_3.

Calculation of the "Wet" Plutonic Facies

Saturated Norm

Ah Ap	Il	Mt_0	Or	Ab	An	Wo	Hy	ΔQ	
(2) *21*	88	37	95	520	397	192	732	−315	
	−46					−46	+46	−23	*Sphene = 69*
	42		95	520	397	146	778	−338	
	−42		−87	−474	−362	−146	−742	+386	*Hastingsite = 1467*
	—		8	46	35	—	36	+48	
			−4				−4	+ 2	*Biotite = 6*
Fsp* = 85			4	46	35		32	+50	*Quartz = 50*
			Or_5	Ab_{54}	An_{41}		−32		*Hypersthene = 32*
				85			—		*Andesine = 85*

	Atoms	Factors	Vol.equ.	Vol.-%	
Quartz	50	1.36	68	3.7	
Andesine	85	1.21	103	5.5	
Hypersthene	32	1.00	32	1.7	
Hastingsite	1467	1.03	1510	81.4	
Biotite	6	1.10	7	0.4	
Sphene	69	1.12	77	4.2	CI = 90.8
Magnetite	37	0.91	34	1.8	
Apatite	21	1.14	24	1.3	
	1767		1855	100.0	
(Cc + Ah = 8)					

Result: *Hornblendite*

Notes

Key 3

step 12: $w = 192/732 = 0.262$, $x = 0.524 \cdot 88 = 46$.
step 17: $k = 95/615 = 0.154$, $w^* = 146/778 = 0.188$,
$\qquad m^* = 532/778 = 0.685$,
$\qquad q'' = -0.52$, $f'' = 1.23$, $i'' = 0.18$.
step 18: $H = (0.188 + 0.06) \times 778/026 = 742$,
$\qquad (Il'' = 133)$, $Il^* = 42$, $Il'' = Il$,
$\qquad D = (1.23 \times 742)/1012 = 0.913$, $Q'' = -0.52 \times 742 = 386$.
step 21: $X = 1.2(8 - 4.6) = 4$.

In the plutonic facies the small amount of SO_3 can be neglected, as the hornblendite is oversaturated and cannot contain haüyne. It is very probable that the SO_3 in the lava has been added to the pyromagma by gaseous transfer.

Calculation of the Mixed Facies

Under plutonic conditions 81.4 vol.-% of hastingsite would have formed, but the process of crystallization was interrupted by a volcanic eruption when only 18% of hastingsite had formed. These 18% must be subtracted from the saturated norm, whereas the remainder will be calculated as formed under volcanic conditions. After having established the mineral assemblages for both the plutonic and the volcanic facies, the mixed facies is calculated as follows:

Reduce the saturated norm of hastingsite, multiplying each component by the factor $18/81.4 = 0.221$. The hastingsite phenocrysts will contain $0.221 \times 1467 = 326$ atoms. All components are reduced proportionally and subtracted member by member from the saturated norm of the rock. The remainder will be distributed according to the volcanic facies (already calculated). Since there is 160 Wo left over, the titanaugite will be reduced proportionally to $160/192 = 0.835$. Then, haüyne will be subtracted as a whole. The remaining $Hy^* = 408$ is converted into 306 olivine and 102 free silica. There remains a silica deficiency of -84 atoms which will be compensated by converting 214 alkaline feldspar into 130 nepheline. The remaining feldspars are determined by the nomogram of Fig. 49 (volcanic facies):

Ah	Ap	Il	Mt₀	Or	Ab	An	Wo	Hy	ΔQ	
2	21	88	37	95	520	397	192	732	−315	
		−9		−20	−107	−81	−32	−164	+87	*Hastingsite* = 326
2		79		75	413	316	*160*	568	−228	= remainder
		−22		−2	−23	−60	−160	−160	+30	*Ti-augite* = 397
2		57		73	390	256	—	408	−192	
2				−4	−26				+12	*Haüyne* = 16
								408	−186	
								−408	+102	*Olivine* = 306
				69	364	256		—	−84	
				−12	−202				+84	*Nepheline* = 130
Fsp* = 475				57	162	256		—		*Sanidine* = 31
6½% San. = 31				Or₁₂	Ab₃₄	An₅₄				*Labradorite* = 444

	Atoms	Factors	Vol.equ.	Vol.-%	
Sanidine	31	1.28	40	2.2	$A = 5.5$
Labradorite	444	1.21	536	29.0	$P = 72.7$ $x = 92.9$
Nepheline	130	1.08	140	7.6	$F = 21.8$ $y = -21.8$
Haüyne	16	1.28	21	1.1	
Titanaugite	397	1.00	397	21.5	
Hastingsite	326	1.03	337	18.2	
Olivine	306	0.88	269	14.5	$CI = 60.1$
Magnetite	37	0.91	34	1.8	
Ilmenite	57	0.92	52	2.8	
Apatite	21	1.14	24	1.3	
	1769		1850	100.0	
(+ sec. Cc = 6)					

Comment

The result corresponds to the mode as far as it concerns the crystalline phases. However, sanidine and nepheline are virtually contained in the glassy to crypto-crystalline matrix. The full name of the rock would be *pheno-basaltic augite hornblende basanite*.

Example No. 9: "Leucite-tephrite"

(Vesuvius/Italy, eruption of 1944; average of 8 analyses;
A. SCHERILLO, 1949, p. 180)

Weight %		Atoms	Ah	Hl	Ap	Il	Mt_0	Or	Ab	An	Wo	Hy	ΔQ	
SiO_2	48.08	800						432	246	132	78	142	-230	
Al_2O_3	18.25	358						144	82	132				
Fe_2O_3	5.06	63												
FeO	4.21	59				9	62					51		
MgO	3.66	91										91		
CaO	8.98	160	1		17					64	78			
Na_2O	2.80	90		8					82					
K_2O	6.79	144						144						
TiO_2	0.70	9				9								
P_2O_5	0.73	10			10									
BaO	0.31	2								2				
Cl	0.27	(8)		(8)									(Cl = 7.6)	
SO_3	0.06	1	1										$(SO_3 = 0.75)$	
		1787	2	8	27	18	62	720	410	330	156	284	-230	
						-2		-9	-16	-42	-156	-156	$+23$	Augite = 358
			2	8		16		711	394	288	—	128	-207	
				-8					-114				$+46$	Sodalite = 76
			2	—				711	280	288		128	-161	
			-2					-3	-19				$+99$	Haüyne = 15
			—					708	261	288		128	-152	
												-128	$+32$	Olivine = 96
								708	261	288		—	-120	
								-545	-55				$+120$	Leucite = 480
Fsp* = 657								163	206	288			—	Sanidine = 174
Sanidine = $26^1/_2$ %								Or_{25}	Ab_{31}	An_{44}				Labradorite
														= 483

Notes

Key 2

step 22: $F'Wo = 0.43 \times 156 = 67$,
$\quad k = 720/1130 = 0.637$,
$\quad n = 1$, $\quad Wo/An = 0.47$, $\quad p = 0.14$,

$x =$	144	9
$y =$	246	16
$z =$	660	42
	1050	67

$\quad Q' = 23$.

step 28: $k^* = 708/968 = 0.732$,
$\quad q^* = 120/968 = 0.124$.

	Atoms	Factor	Vol.equ.	Vol.-%	
Sanidine	174	1.28	223	10.6	$A = 14.3$
Labradorite	483	1.21	584	27.7	$P = 37.4$ $x = 72.5$
Leucite	480	1.33	638	30.3	$F = 48.3$ $y = -48.3$
Sodalite	76	1.28	97	4.6	
Haüyne	15	1.28	19	0.9	
Augite	358	1.00	358	17.0	
Olivine	96	0.88	84	4.0	$CI = 25.9$
Ti-magnetite	78	0.91	71	3.4	
Apatite	27	1.14	31	1.5	Field 13
(control)	1787		2105	100.0	

Result: *Leucite phonotephrite*

Comment

Sodalite and haüyne have not been observed, and also the presence of sanidine is questionable even in specimens with very little glass. One may suppose that Cl and SO_3 are dissolved in the glass and will be lost if the rock crystallises completely. If so, Cl and SO_3 should be neglected.

Atoms	Ap	Il	Mt$_0$	Or	Ab	An	Wo	Hy	ΔQ	
Si 800				432	270	124	83	142	-251	
Al 358				144	90	124				
Fe 122		9	62					51		
Mg 91								91		
Ca 162	17					62	83			(incl. Ba)
Na 90					90					
K 144				144						
Ti 9		9								
P 10	10									
1786	27	18	62	720	450	310	166	284	-251	
		-3		-10	-19	-43	-166	-166	$+25$	*Di-augite* $= 382$
		15		710	431	267	—	118	-226	
								-118	$+30$	*Olivine* $= 88$
				710	431	267		-196		
				-614	-61			$+135$		*Leucite* $= 540$
				96	370	267		-61		
				-32	-120			$+61$		*Nepheline* $= 91$
Fsp* $= 581$				64	250	267		—		*Sanidine* $= 29$
Sanidine $= 5\%$				Or$_{11}$	Ab$_{43}$	An$_{46}$				*Plagioclase* $= 552$

	Atoms	Factor	Vol.equ.	Vol.-%	
Sanidine	29	1.28	37	1.8	$A = 2.5$
Plagioclase	552	1.21	668	32.1	$P = 43.9$ $x = 92.6$
Leucite	540	1.33	718	34.5	$F = 53.6$ $y = -53.6$
Nepheline	91	1.08	98	4.7	
Diopside-augite	382	1.00	382	18.3	
Olivine	88	0.88	77	3.7	$CI = 26.9$
Ti-magnetite	77	0.91	70	3.4	field 14
Apatite	27	1.14	31	1.5	
	1786		2081	100.0	

The result of this calculation differs considerably from the previous one. Leucite and plagioclase are more abundant, whereas sanidine is nearly nil. The projecting point in the double-triangle has shifted from field 13 into field 14: = *leucite tephrite*. This agrees much better with the mode, thus confirming the probably accidental character of Cl and SO_3.

Among the numerous ejected blocks of Somma-Vesuvius, dike rocks are found containing some biotite, besides leucite, plagioclase, etc. They represent the subvolcanic facies of leucite tephrites. In order to calculate the mineral assemblage of this heteromorphic rock, the same Key 2 is used, proceeding, however, from step 24 to step 25. This variation of the trend of calculating is based on the assumption of a relatively high vapour pressure, causing the formation of biotite under subvolcanic conditions.

Ap	Il	Mt_0	Or	Ab	An	Wo	Hy	ΔQ	
27	18	62	720	450	310	166	284	−251	sum = 1786
	−3		−10	−19	−43	−166	−166	+25	*Diopside-augite* = 382 + 10
	15		710	431	267	—	118	−226	
	−9		−90	−12	−11	+5	−113	+59	*Biotite* = 171
	6		620	419	256	(5)	(5)	−167	
			−527	−53				+116	*Leucite* = 464
			93	366	256	—	—	−51	
			−27	−100				+51	*Nepheline* = 76
Fsp* = 588			66	266	256			—	*Sanidine* = 12
			Or_{11}	Ab_{45}	An_{44}				*Plagioclase* = 576

	Atoms	Factor	Vol.equ.	Vol.-%	
Sanidine	12	1.28	15	0.7	$A = 1.0$
Plagioclase	576	1.21	697	33.5	$P = 49.5$ $x = 98.0$
Leucite	464	1.33	617	29.6	$F = 49.5$ $y = -49.5$
Nepheline	76	1.08	82	3.9	
Di-augite	392	1.00	392	18.8	
Biotite	171	1.10	188	9.0	$CI = 32.3$
Ti-magnetite	68	0.91	62	3.0	
Apatite	27	1.14	31	1.5	field 14
	1784		2084	100.0	

Similar rocks are rather common among the ejected blocks of Mount Somma. Locally, these "leucite theralites" are called sommaites; it has to be mentioned that there are also leucite tephrites which contain small amounts of biotite. They present the characteristics of mixed facies.

Other types of ejected blocks contain melanite and amphibole, besides nepheline and feldspars. The following calculation of the plutonic facies reveals them as heteromorphs of leucite tephrites.

Ap	Il	Mt_0	Or	Ab	An	Wo	Hy	ΔQ	
27	18	62	720	450	310	166	284	-251	sum $= 1786$
	-15				-15	-120	-45	$+30$	$Melanite = 165$
	3		720	450	295	46	239	-221	
	-3		-114	-71	-49	-46	-232	$+97$	$Hastingsite = 418$
	—		606	379	246	—	7	-124	
			-5	-1	-1		-7	$+4$	$Biotite = 10$
			601	378	239		—	-120	
			-45	-255				$+120$	$Nepheline = 180$
Fsp* $= 924$			556	123	245			—	$Orthoclase = 610$
			Or_{60}	Ab_{13}	An_{27}				$Bytownite = 314$

Notes
Key 3

step 1: Alternative b).

step 12: $w = 166/284 = 0.584$, $X = 2 \times 0.416 \times 18 = 15$.

step 17: $w* = 46/239 = 0.192$, $m = 91/142 = 0.64$,
     $w_1'' = 0.2$, $i'' = 0.14$, $f'' = 1.01$, $q'' = -0.42$.

step 18: $H = (0.192 + 0.060) \times 239/0.26 = 232$,
     $D = (1.01 \times 232)/1480 = 0.158$.

step 21: $X = 1.2 \times (606 - 38) = 682$.
step 22: $k' = 0.3 \times 0.615 + 0.70 = 0.885$.
step 36: $k^* = 601/979 = 0.614$,
$\qquad q^* = 120/979 = 0.123, \quad U = 1.5$.
step 37: $X = 2.5 \times 120 = 300$.

Key 6

	Atoms	Factor	Vol.equ.	Vol.-%		
Orthoclase	610	1.28	781	38.2	$A = 57.7$	
Bytownite	314	1.21	380	18.6	$P = 28.0$	$x = 32.8$
Nepheline	180	1.08	194	9.5	$F = 14.3$	$y = -14.3$
Hastingsite	418	1.03	431	21.0		
Biotite	10	1.10	11	0.5	$CI = 33.7$	
Melanite	165	0.99	163	8.0		
Magnetite	62	0.91	56	2.7	Field 12	
Apatite	27	1.14	31	1.5		
	1786		2047	100.0		

Result: *Melanite-bearing amphibole foid monzosyenite*

Comment

   Very similar rocks are known among the ejected blocks of Mt. Somma-Vesuvius.

Example No. 10: "Nephélinite mélilitique à olivine et kalsilite"

(Nyiragongo/East Africa; TH. G. SAHAMA [in: M.-E. DENAYER and F. SCHELLINCK 1965, p. 94/95, no. F 6 (1)])

Weight %		Atoms	Cc	Ap	Il	Or	Ab	Ac	Mt_0	Wo	Hy	ΔQ
SiO_2	36.56	609				369	387	90		186	175	−598
Al_2O_3	12.85	252				123	129					
Fe_2O_3	4.80	60						45				
FeO	7.10	99			39				31		48	
MnO	0.28	4										
MgO	5.12	127									127	
CaO	14.56	260	19	55						186		
Na_2O	5.38	174					129	45				
K_2O	5.80	123				123						
TiO_2	3.13	39			39							
P_2O_5	2.31	33		33								
CO_2	0.85	19	19									
		1799	*38*	*88*	78	615	645	180	*31*	372	350	−598
					−14	−14	−112	−180		−372	−350	+45 *Aegirinaugite* = 997
					64	601	533	—		—	—	−553
						−550						+110 *Leucite* = 440
							584					−443
$Mt_0 + Il'' = 31 + 39 = 70$							−584					+234 *Nepheline* = 350
						—						−209 $= \Delta Q_m$

Il″	(Or′ + Ab′)	Ac	Wo	Hy	Q′	Provisional
39	126	180	372	350	−45	*Clinopyroxene* $Cpx_0 = 1022$
−39			−39	+39		*Perovskite* $Psk_0 = 39$
		−180	−90		+45	
	−109		−243	−102	+100	*Melilite* $Mel_0 = 579$
				−287	+72	*Olivine* $Ol_0 = 215$
	−17				+7	*Nepheline* $Ne_0 = 10$
—	—	—	—	—	179	$= \Sigma Q_m$

R = 1.168   X = 220
R* = 0.757   (1 − R*) = 0.243

		Atoms	Factor	Vol.equ.	Vol.-%	
Leucite	$440 - 1.333\,X$	146	1.33	194	9.9	
Kalsilite	X	220	1.17	257	13.1	$F = 100$
Nepheline	$10\,R^* + 350$	358	1.08	387	19.8	$y = -100$
Aegirine-augite	$(1 - R^*)\cdot 1022$	248	1.00	248	12.7	
Olivine	$215\,R^*$	163	0.88	143	7.3	
Melilite	$579\,R^*$	438	1.12	491	25.4	
Ulvöspinel	$39 + 31$	70	0.91	64	3.3	$CI = 57.2$
Perovskite	$39\,R^*$	30	1.02	31	1.6	
Apatite		88	1.14	100	5.1	
Calcite		38	1.11	42	2.1	
	Control: 1799			1957	100.0	

Result: *Leucite-bearing melilite nephelinite*

Notes
Key 1

step 2: $Ox^0 = 60/163 = 0.368$.
step 3: $\tau = (252 - 174)/39 = 2.00$.
         $\sigma = (5.38 + 580)^2/(36 \cdot 56 - 43.00) = -19.4$.
step 4: Calcite $= 2 \cdot 19 = 38$; remaining Ca $= 260 - 19$.
step 5: Alternative c).
step 8: Mt_0 is formed after Ac, $Mt_0 = 0.397\,(163 - 39 - 45) = 31$.

Key 2

step 1: Alternative c).
step 18: Alternative b).
step 23: $M = 902$.
step 26: Alternative a).
step 27: $Hy^* = 0$,   alternative c): $\Delta Q^* = \Delta Q_1$.
step 28: $k = 615/1260 = 0.488$,   alternative b),
         $k^* = 601/1134 = 0.530$,
         $q^* = 553/1134 = 0.488$,
         $v' = 1260/722 = 1.745$ (Fig. 47 → Fig. 48).
step 36: $q_L = 0.097$, $X = 0.097\,(601 + 533) = 110$.
step 40: Alternative c).
step 43: $Il'' = 0.5\,Il = 39$.
step 44: Alternative c), $R = 209/179 = 1.168$.
step 46: Alternative b).
step 48: $X = (1.168 - 0.3)\times(209 + 45) = 220$.
         Alternative a),
         $R^* = (209 - 0.333 \times 220)/179 = 0.757$,
         $(1 - R^*) = 0.243$.

Comment

The result agrees fairly well with the observed mode, which is rather variable from slide to slide, as is usual in melilite-bearing rocks.

The low τ-value indicates the common parentship with upper mantle basalts, whereas the negative value of σ reflects the strong desilication and the alkaline serial character.

The heteromorphic equivalent of melitite-bearing rocks are carbonatitic rocks to be calculated with the aid of Key 5.

Calculation of the Heteromorphic Carbonatite

Cc	Ap	Il	Or	Ab	Ac	Mt_0	Wo	Hy	ΔQ	
38	*88*	78	615	645	180	31	372	350	-598	
		-14	-14	-112	-180		-372	-350	$+45$	*Aegirinaugite* $=997$
		64	601	533	—		—	—	-553	
			-85	-483					$+227$	*Nepheline* $=341$
			516	50					-326	
			-516	-50						*Orthoclase* $=566$
			—	—					-326	$=\Delta Q_C$

Provisional ilmenite: $Il_0 = 14$,
Provisional orthoclase: $Ort_0 = 1.2 \times 14 = 17$,
Provisional nepheline: $Ne_0 = 0.6 \times (112 - 3) = 65$,
Provisional carbonates: $Carb_0 = 372 + 350 = 722$,
$\Sigma Q_C = 361 + 43 - 45 = 359$,
$R = 326/359 = 0.90$.

		Atoms	Factor	Vol. equ.	Vol.-%		
Orthoclase	$566 + 17\,R$	581	1.28	744	31.0	$A =$	63.3
Nepheline	$341 + 65\,R$	400	1.08	432	18.0	$F =$	36.7
Aegirine	$997\,(1 - R) + 80\,R$	256	1.00	256	10.6	$x =$	0
Ulvöspinel	$31 + 64 + 14\,R$	108	0.91	98	4.1	$y =$	-36
Apatite		88	1.14	100	4.2	$CI =$	18.9
Carbonates	$38 + 722\,R$	694	1.11	770	32.1	Carb $=$	32
		2127		2400	100.0		
		-328					
	Control: 1799				$CI' = 1890/67.9 = 27.8$		

Result: *Carbonatite foyaite*

Example No. 11: "Melilitith"

(Homboll, Hegau/SW-Germany; W. von ENGELHARDT
and W. WEISKIRCHNER, 1963, p. 22, no. 9)

weight %	atoms	Ap	Il	Mt$_0$	Or	Ab	An	Wo	Hy	ΔQ	
SiO$_2$ 37.19	619				147	291	47	196	474	−536	
Al$_2$O$_3$ 9.83	193				49	97	47				
Fe$_2$O$_3$ 8.31	104										
Cr$_2$O$_3$ 0.06	1		24	45							
FeO 4.88	68								106		
MnO 0.15	2										
MgO 14.84	368								368		
CaO 13.24	236	17					23	196			
Na$_2$O 3.02	97					97					
K$_2$O 2.32	49				49						
TiO$_2$ 1.94	24		24								
P$_2$O$_5$ 0.68	10	10									
	1771	27	48	45	245	485	117	392	948	−536	
			−48		−25	−146	−117	−392	−392	+101	*Augite* = 1019
			—		220	339	—	—	556	−435	
									−556	+139	*Olivine* = 417
					220 + 339			—		−296	
					−370					+148	*Nepheline* = 222
					189					−148	
					−189					+38	*Leucite* = 151
									—	−110	= ΔQ$_m$

In order to compensate the silica deficiency ΔQ$_m$ some Cpx must be converted into melilite etc. Before doing so, $^1/_2$Il = Il″ is added to Mt$_0$, the other half remaining in the provisional Cpx$_0$.

Il″	Or′ + Ab′	An	Wo	Hy′	Q′	
24	171	117	392	392	− 101	$Cpx_0 = 995$
− 24		− 60			+ 24	$Psk_0 = 24$
						$Hz_0 = 36$
—	171	57	392	392	—	
		− 57	− 23		+ 23	$Geh_0 = 57$ $\left.\right\}$ = 596
	− 166		− 369	− 155	+ 151	$Mel_0 = 539$
	5			237		
				− 237	+ 59	$Ol_0 = 178$
	5					
	− 5				+ 2	$Ne_0 = 3$
	—				+ 158	$= \Sigma Q_m$

R = 110/158 = 0.696   R′ = (110 − 37.8)/158 = 0.457
                                       (1 − R′) = 0.543

		Atoms	Factor	Vol.equ.	Vol.-%		
Kalsilite		113	1.17	132	7.4		
Nepheline	0.457 × 3 + 222	223	1.08	241	13.6	F =	100.0
Augite	0.543 × 995	541	1.00	541	30.5	y =	− 100
Olivine	0.457 × 178 + 417	499	0.88	439	24.7		
Melilite	0.457 × 596	272	1.12	305	17.2	CI =	79.0
Perovskite	0.457 × 24	11	1.02	11	0.6		
Ti-magnetite	0.457 × 36 + 45 + 24	85	0.91	77	4.3		
Apatite		27	1.14	31	1.7		
		1771		1777	100.0		

Result: *Mela olivine melilite nephelinite*

Notes
Key 2

step 22: $x = 49$ | 25   F′Wo = 0.736 × 392 = 288
           $y = 290$ | 146
           $z = 234$ | 117   Q′ = 0.35 × 288 = 101
          _____|_____
           573 | 288
step 28: $k^* = 220/559 = 0.394$, $q^* = 296/559 = 0.530$.
step 33: $q_N = 0.265$, X = 148 .
step 44: Kalsilite = 0.75 × 151 = 113 .

Comment

In melilite-bearing rocks, the mode often varies within the same specimen. Furthermore the distinction of kalsilite from nepheline is optically mostly impossible. Also, the distinction of small grains of melilite from augite is sometimes difficult. Therefore the determination of the mode is often not reliable.

Under subvolcanic conditions with higher CO_2 pressure, melilite is not stable. At its place, carbonates will form in addition to silicates. The heteromorphic equivalents of melilite-bearing rocks are carbonatites which can be calculated according to Key 5, as follows:

Ap	Il	Mt_0	Or	Ab	An	Wo	Hy	ΔQ	(saturated norm)	
27	48	45	245	485	117	392	948	−536		
	−36		−25	−146	−117	−392	−392	+101	*Augite*	= 1007
	12		220	339	—	—	556	−435		
	−12		−176	−30			−208	+107	*Biotite*	= 319
	—		44	309			348	−328		
							−348	+87	*Olivine*	= 261
			44	309			—	−241		
			−44	−309				+141	*Nepheline*	= 212
			—	—				−100	$= \Delta Q_c$	

Total conversion of augite into carbonates and silicates:

Provisional Cpx_0 = 1007
Provisional Il_0 = 36
Provisional Ort_0 = 30
Provisional Ne_0 = 85
Provisional Sp_0 = 70
Provisional $Carb_0$ = 784

$Q' = 101$

$\Sigma Q_c = 392 + 103 = 394$.

$R = 100/394 = 0.254$.

$(1 - R) = 0.746$.

		Atoms	Factor	Vol.equ.	Vol.-%
Orthoclase	$R \cdot 30$	8	1.28	10	0.5
Nepheline	$R_c \cdot 85 + 212$	234	1.08	253	13.2
Augite	$(1 - R_c) \cdot 1007$	751	1.00	751	39.3
Biotite		319	1.10	351	18.3
Olivine		261	0.88	230	12.0
Ore	$R_c \cdot 106 + 45$	72	0.91	66	3.5
Apatite		27	1.14	31	1.6
Carbonates	$R_c \cdot 784$	199	1.11	221	11.6
		1871		1913	100.0
	$-CO_2$	−100			
	Control:	1771			

Result: *Carbonatitic melteigite*

Key 5

step 6: Cpx_0 equals that of the volcanic facies minus the amount of Il entering biotite.

step 9: Biotite contains 12 Il which are subtracted from Il' because after the formation of Cpx_0 no Il* has been left over. As $En/Hy = 0.80$, the biotite can said to be phlogopite.

step 13: Note that the conversion of An' is based on the equation:
5 An + 2 Hy + 1 CO_2 = 3 Sp + 3 Q + 2 Cc. However, as Cc is added to the carbonates, and the corresponding amount is subtracted from Hy, the scheme is simplified:
$Sp_0 = 0.6$ An' and *carbonates* = Wo + Hy; $CO_2 = \frac{1}{2}$ carb.

Comment

Similar carbonatitic melteigites are known among inclusions and ejected blocks.

Example No. 12: "Venanzite"

(Leucite kalsilite melilitite; San Venanzo/Italy;
M. MITTEMPERGHER, 1965, p. 460)

Weight %	atoms	Cc	Ap	Il	Mt_0	Or	Ab	An	Wo	Hy	ΔQ
SiO_2 40.52	675					471	108	12	227	364	−507
Al_2O_3 10.43	205					157	36	12			
Fe_2O_3 4.97[13]	62										
FeO 2.92	41			9	45					50	
MnO 0.11	1										
MgO 12.65	314									314	
CaO 16.23	289	48	8					6	227		
Na_2O 1.11	36						36				
K_2O 7.41	157					157					
TiO_2 0.74	9			9							
P_2O_5 0.32	5		5								
CO_2 2.11	48	48									
	1842	96	13	18	45	785	180	30	454	728	−507

Saturated norm:	Il	Or	Ab	An	Wo	Hy	ΔQ	
	18	785	180	30	454	728	−507	
	−18	−155	−107	−30	−454	−454	+102	*Diopside-augite* = 1116
	—	630	73	—	—	274	−405	
						−274	+68	*Olivine* = 206
		630	73			—	−337	
		−623	−62				+137	*Leucite* = 548
		7	11				−200	
		−7	−11				+7	*Nepheline* = 11
		—	—				−193	= $ΔQ_m$

13 Including 0.31 Cr_2O_3.

Note: The calculation of the clinopyroxene is carried out as usual, but the product $F'Wo = 0.707 \times 454 = 321$ leads to a value of An' greater than An, i.e. $59 > 30$. Hence, $F'Wo > Fsp' = 292$, and $Q' = 0.35 \times 292 = 102$.
The silica deficiency $\Delta Q_m = -193$ must be compensated by converting augite into melilite, etc. as follows:

Provisional clinopyroxene: (step 39) $Il'' = 0.5 \, Il$

Il''	$Or' + Ab'$	An	Wo	Hy'	Q'			
9	262	30	454	454	−102	Prov. $Cpx_0 = 1107$		
−9		−22			+9	Prov. $Psk_0 = \quad 9$	and	$Hz_0 = 13$
		−8	−4		+4	Prov. $Mel_0 = \quad 665$		
	−203		−450	−189	+185			
				−265	+66	Prov. $Ol_0 = \quad 199$		
	−59				+24	Prov. $Ne_0 = \quad 35$		
—	—	—	—	—	186	$= \Sigma Q_m$		

step 44c: $R = 193/186 = 1.0376$.
step 48: $X = 0.7376 \times 298 = 220 < 0.75 \, Lc$,
$\qquad\quad X = $ Kalsilite,
$\qquad\quad R^* = (193 - 73.3)/186 = 0.6435$.

		Atoms	Factor	Vol. equ.	Vol.-%	
Leucite	$548 - 293$	255	1.33	339	17.1	
Kalsilite		220	1.17	257	13.0	$F = 100$
Nepheline	$23 + 11$	34	1.08	37	1.9	
Diopside-augite	0.356×1107	394	1.00	394	19.9	
Olivine	$128 + 206$	334	0.88	294	14.8	
Melilite	$R^* \cdot 665$	428	1.12	479	24.1	$CI = 68.0$
Ore	$8 + 9 + 45$	62	0.91	56	2.8	
Perovskite	$R^* \cdot 9$	6	1.02	6	0.3	
Apatite		13	1.14	15	0.7	
Calcite		96	1.11	107	5.4	
Control		1842		1984	100.0	

Result: *Kalsilite leucite melilitite*

Comment

The result corresponds to the observed mode of the normal variety of "venanzite" which, like most melilitites, is rather inhomogeneous presenting various modes within the same flow.
Some varieties of "venanzite" contain phlogopite and greater amounts of carbonates, presenting, thus, mixed facies between volcanic melilitite and subvolcanic carbonatite. This latter can be calculated by the aid of Key 5 following steps $1a - 2b - 6 - 8a - 9a - 10a - 11 - 12a - 13 - 15$.

Note that step 6 (clinopyroxene) equals step 22 of Key 2 which has already been calculated and subtracted from the saturated norm (see p. 129). The remainder has been:

Or*	Ab*	Hy*	ΔQ*		
630	73	274	-405	Diopside-augite	$= 1116$
-232	-14	-274	$+142$	Phlogopite	$= 378$
398	59	—	-263		
-3	-20		$+9$	Nepheline	$= 14$
395	39		-254		
-395	-39		0	Orthoclase	$= 434$
—	—		-254	$= \Delta Q_c$	

Total conversion of the diopside-augite (Cpx_0) into provisional

$$
\begin{aligned}
Il_0 & & &= 18 \\
Ort_0 &= 1.2 \times 155 & &= 186 \\
Ne_0 &= 0.6 \,(107 - 31) & &= 45.6 \\
Sp_0 &= 0.6 \times 30 & &= 18 \\
Carb_0 &= 454 + 454 & &= 908
\end{aligned}
$$

$$\Sigma Q_c = 0.5 \cdot 908 + 0.667 \,(45.6 + 18) - 102 = 394.4$$

$$R_c = 254/394.4 = 0.644.$$

According to step 15 of Key 5 and the conversion in volume percent by the aid of Key 6, one obtains the following result:

		Atoms	Factor	Vol.equ.	Vol.-%
Orthoclase	$434 + R_c \cdot 186$	554	1.28	709	29.5
Nepheline	$14 + R_c \cdot 45.6$	43	1.08	46	1.9
Diopside-augite	$(1 - R_c) \cdot 1116$	397	1.00	397	16.6
Phlogopite		378	1.10	416	17.3
Ulvöspinel	$45 + R_c \cdot 36$	68	0.91	62	2.6
Apatite		13	1.14	15	0.6
Carbonates	$96 + R_c \cdot 908$	681	1.11	756	31.5
		2134		2401	100.0
Minus added CO_2		-292			
Control		1842			

Result: *Carbonatitic phlogopite augite alkali syenite*

Comment

Similar rocks may be found as ejected blocks. Some varieties of so-called venanzite are typical mixed facies between the above carbonatitic alkali syenite and the normal melilititic lava.

Appendix

1. Key for Calculation of Strongly Altered Rocks (Sil)

A) Calculate: $Z = 0.33\,(1 - h)$ and distinguish two alternatives:
a) $q > Z$:

Sil	Hy	ΔQ	
− Sil*	− Hy	− 0.33 − ΔQ*	*Hypersthene* = Hy *Kaolinite* = sum *Quartz* = ΔQ*
—	—	—	────────→ Key 2 ㊳

b) $q < Z$: Calculate: $X = 0.5\,\text{Sil} - 1.5\,\Delta Q$

Or	Ab	Sil	Hy	ΔQ	
− 1.54	− 0.31	− **X** − **Sil***	− Hy	+ 0.33 − 0.33	*Hypersthene* = Hy *Muscovite* = sum *Kaolinite* = sum
Or*	Ab*	—	—	(ΔQ*)	────────→ Key 2 ㊳

If $\Delta Q^* < 0$, then convert *hypersthene* into *olivine* according to the relation: $4\,\text{Hy} - 1\,\text{Q} = 3\,\text{Ol}$.

B) Calculate: $Z = 0.33 - 0.915\,h$ and $k' = 0.3\,k + 0.70$ and distinguish two alternatives:
a) $q > Z$:

Il	Or	Ab	Sil	Hy	ΔQ	
− 0.1	− 0.9 k'	− 0.9 $(1 - k')$	− 0.22 − **Sil***	− **Hy**	+ 0.52 − 0.33 − ΔQ*	*Biotite* = sum *Kaolinite* = sum *Quartz*
Il*	Or*	Ab*	—	—	—	──→ Key 2 ㊳

b) $q < Z$:   Calculate: $X = 0.5 \, Sil - 1.5 \, \Delta Q - 0.89 \, Hy$
and $k' = 0.3 \, k + 0.70$

Il	Or	Ab	Sil	Hy	ΔQ		
-0.1	$-0.9 \, k'$	$-0.9 \, (1-k')$	-0.22	$-$ **Hy**	$+0.52$	*Biotite*	$=$ sum
	-1.54	-0.31	$-$ **X**		$+0.33$	*Muscovite*	$=$ sum
			$-$ **Sil***		-0.33	*Kaolinite*	$=$ sum
Il*	Or*	Ab*	$-$	$-$	$(\Delta Q^*) \longrightarrow$ Key 2	㉟	

If $\Delta Q < 0$, then convert: $4 \, Hy - 1 \, Q = 3 \, Ol$.

2. Keys for Calculation of the Mineral Assemblages of Ultramafic Rocks

Among ultramafic rocks only hornblendites and eclogites correspond chemically to basaltic or tephritic magmas which have consolidated under plutonic or high pressure conditions. Peridotites, pyroxenites and similar rocks may be cumulites, metamorphites, residual products of degranitisation or upper mantle material, the chemical composition of which does not correspond to any known magma erupting at the earth's surface. The calculation of the mineral assemblage of cumulites of magmatic origin cannot be carried out by the aid of the Keys 3 or 4, because the composition of the magma from which they settled is mostly unknown. Hence, also the composition of the accumulated phenocrysts appears to be rather problematic.

In order to calculate approximately the probable mineral assemblages of ultramafic rocks, particular keys are given on the following pages. Also here, several facies have to be distinguished:

A. The "dry" plutonic facies.

B. The "wet" plutonic facies.

C. A high pressure or eclogite facies.

In each one of these facies some characteristic minerals will appear:

In A: Clinopyroxenes, orthopyroxenes and occasionally olivine.

In B: Amphiboles and biotite (phlogopite), besides pyroxenes.

In C: Garnets and pyroxenes (omphacite), occasional orthopyroxene, olivine, etc.

Mixed facies are frequent; therefore it must be interpolated between two of the mentioned facies.

*Key A: Probable Mineral Assemblages of Ultramafic Rocks
in the "Dry" Plutonic Facies*

① Distinguish two alternatives:

a) the saturated norm contains *An* ——————————————→ ②

b) the saturated norm contains *Ac* ——————————————→ ⑤

② Distinguish two alternatives:

a) Or > 0.1 Ab
  then: $Or_0 = 1.2$ Or [resp. (Or + Ab) if Ab $\leqq$ 0.2 Or]
  $Ab_0 = Ab - 0.2$ Or ——————————————→ ③

b) Or $\leqq$ 0.1 Ab
  then: $Ab_0 = Ab + Or$ ——————————————→ ③

③ Distinguish two alternatives:

a) Wo > 1.5 $(Ab_0 + An)$ ——————————————————→ ④

b) Wo $\leqq$ 1.5 $(Ab_0 + An)$
  Calculate: X = 0.8 $(Ab_0 + An - 0.67$ Wo)
  and replace: An by An − X,
        Wo by Wo + 0.4 X,
        Sil by Sil + 0.6 X ——————————————→ ④

④ Distinguish two alternatives:

a) Wo > 1.5 $(Ab_0 + An)$
Calculate: Q′ = 0.35 $(Ab_0 + An)$

Ab_0	An	Wo	Hy	ΔQ	
$-Ab_0$	− An	− Wo	− 1.5 Wo	+ Q′	*Clinopyroxene*
—	—	—	Hy*	ΔQ*	——→ ⑥

b) Wo $\leqq$ 1.5 $(Ab_0 + An)$

Ab_0	An	Wo	Hy	ΔQ	
	− 0.67	**− Wo**	− 1.5	+ 0.23	*Clinopyroxene*
Ab*	An*	—	Hy*	ΔQ*	——→ ⑥

⑤ Calculate: X = 0.1 (Wo + 0.5 Ac)

Ab	Ac	Wo	Hy	ΔQ	
− 3 X	− Ac	− Wo	− Wo	+ X	*Clinopyroxene (alkaline)*
Ab*	—	—	Hy_1	$ΔQ_1$	——→ ⑫

⑥ Distinguish two alternatives:
   a) Sil = nil ————————————————————————————→ ⑫
   b) Sil > 0,   then determine the number of the next step
                  according to the following scheme:

	1.5 Hy > Sil	1.5 Hy ≤ Sil
$\Delta Q^* > 0$	⑦	
$0 > \Delta Q^* > -Hy^*$	⑧	
$-Hy^* > \Delta Q^* > -0.67\,Sil$	⑨	⑩
$\Delta Q^* < -0.67\,Sil$		⑪

⑦ $\Delta Q^* = Quartz$
    $Hy^* = Orthopyroxene$
    $Sil\ \ \ = Kyanite$ ————————————————————————————→ ⑬

⑧
Sil	Hy*	ΔQ^*	
-1.5	-1.0	$+\Delta Q^*$	Spinel
$-Sil^*$			Kyanite
	$-Hy^{**}$		Orthopyroxene
—	—	—	————————————→ ⑬

⑨
Sil	Hy*	ΔQ^*	
$-Sil$	-0.67	$+0.67$	Spinel
—	Hy_1	ΔQ_1	————————————→ ⑫

⑩
Sil	Hy*	ΔQ^*	
-1.5	$-Hy^*$	$+1.0$	Spinel
-3.0		$+\Delta Q^{**}$	Corundum
$-Sil^*$			Kyanite
—	—	—	————————————→ ⑬

⑪
Sil	Hy*	ΔQ^*	
-1.5	$-Hy^*$	$+1.0$	Spinel
$-Sil^*$		$+0.33$	Corundum
—	—	ΔQ_2	————————————→ ⑬

⑫ Distinguish two alternatives:
a) $0 > \Delta Q_1 > -0.25\, Hy_1$

Hy_1	ΔQ_1	
-4.0	$+\Delta Q_1$	*Olivine*
$-Hy_1^*$		*Orthopyroxene*
—	—	→ ⑬

b) $\Delta Q_1 < -0.25\, Hy^*$

Hy_1	ΔQ_1	
$-\mathbf{Hy_1}$	$+0.25$	*Olivine*
—	ΔQ_2	→ ⑬

⑬ Distinguish three alternatives:
a) $\Delta Q_2 > 0$ ———————————————→ ⑭
b) $0 > \Delta Q_2 > -0.4\, Ab^*$ ——————→ ⑮
c) $\Delta Q_2 < -0.4\, Ab^*$ ———————————→ ⑯

⑭ Distinguish two alternatives:
a) $An^* = nil$
  $Ab^* = Albite$ ——————————————→ ⑰
b) $An^* > 0$
  $Ab^* + An^* = Plagioclase$ —————————→ ⑰

⑮ Distinguish two alternatives:
a) $An^* = nil$

Ab^*	ΔQ_2	
-2.5	$+\Delta Q_2$	*Nepheline*
$-Ab^{**}$		*Albite*
—	—	→ ⑰

b) $An^* > 0$

Ab^*	An^*	ΔQ_2	
-2.5		$+\Delta Q_2$	*Nepheline*
$-Ab^{**}$	$-An^*$		*Plagioclase*
—	—	—	→ ⑰

⑯ Distinguish two alternatives:
a) $An^* = nil$

Ab^*	ΔQ_2	
$-\mathbf{Ab^*}$	$+0.4$	*Nepheline*
—	ΔQ_3	→ ⑰

b) An* > 0

Ab*	An*	ΔQ_2	
—**Ab***		+0.4	*Nepheline*
	—**An***		*Plagioclase*
—	—	ΔQ_3	⟶ (17)

(17) Distinguish three alternatives: (only if Or* has been formed)

a) $\Delta Q_3 =$ nil
   Or* = *Orthoclase*

b) $0 > \Delta Q_3 > -0.4$ Or*

Or*	ΔQ_3	
—2.5	+**ΔQ_3**	*Kalsilite*
—Or**		*Orthoclase*

c) $\Delta Q_3 < -0.4$ Or*

Or*	ΔQ_3	
—**Or***	+0.4	*Kalsilite*
—	ΔQ_r	⟶ (18)

(18) ΔQ_r will be quoted as such, its compensation being questionable. This extreme silica deficiency may be due to the presence of carbonates or oxides in the mode and/or incomplete or erroneous analysis (e.g. CO_2 not determined). Then enter into Key 6 (p. 131).

Key B: Probable Mineral Assemblages of Ultramafic Rocks in the "Wet" Plutonic Facies

(1) Dinstinguish two alternatives:

a) Or > 0.1 Ab
   $Or_0 = 1.2$ Or (resp. Or + Ab, if Ab < 0.2 Or)
   $Ab_0 =$ Ab — 0.2 Or ⟶ (2)

b) Or ≤ 0.1 Ab
   $Ab_0 =$ Or + Ab ⟶ (2)

(2) Distinguish two alternatives:

a) the saturated norm contains *An* ⟶ (3)

b) the saturated norm contains *Ac* ⟶ (4)

(3) Distinguish two alternatives:

a) 2.5 Wo > $(Ab_0 +$ An) ⟶ (5)

b) 2.5 Wo ≤ $(Ab_0 +$ An)
   Calculate: X = 0.5 $(Ab_0 +$ An — 2.5 Wo)
   and replace: An by An — X
   Sil by Sil + 0.6 X
   Wo by Wo + 0.4 X ⟶ (5)

④	Ac	Hy	ΔQ		
—Ac		-0.75	-0.125	*Riebeckite* (adds *amphibole*)	
—		Hy*	ΔQ*	——————————————→	⑤

⑤ Distinguish two and three alternatives (being Hy < 1.2 Or_0):

    a) $Or_0 =$ nil ————————————————————————————→ ⑨

    b) $Or_0 > 0$

      ') Sil = nil ——————————————————————————→ ⑥

      '') $0 <$ Sil ≤ 0.25 Or_0 ——————————————————→ ⑦

      ''') Sil > 0.25 Or_0 ————————————————————→ ⑧

⑥	Or_0	Hy	ΔQ	(resp. Hy* and ΔQ* after riebeckite)	
—Or_0		-1.2	$+0.6$	*Biotite*	
—		Hy*	ΔQ*	——————————————→	⑨

⑦	Or_0	Sil	Hy	ΔQ		
—Or_0		$-$Sil	-1.15	$+0.58$	*Biotite*	
—		—	Hy*	ΔQ*	——————————→	⑨

⑧	Or_0	Sil	Hy	ΔQ		
—Or_0		-0.25	-1.10	$+0.55$	*Biotite*	
—		Sil*	Hy*	ΔQ*	——————————→	⑨

⑨ Calculate: $X = 0.364$ Hy $- 0.418$ Wo (limits: $X = 0$ and $X =$ Wo),
         $Y = 2X + 0.5$ Wo.

Determine the number of the next step according to the scheme:

	$Y > (Ab_0 + An)$	$Y \leq (Ab_0 + An)$
$X \leq 0$	⑩	⑪
Wo $> X > 0$	⑫	⑬
$X \geq$ Wo	⑭	⑮

⑩ Calculate: $Q' = 0.35\,(Ab_0 + An)$

Ab_0	An	Wo	Hy	ΔQ	(resp. Hy* and ΔQ*)
$-Ab_0$	$-An$	$-Wo$	$-Hy$	$+Q'$	*Clinopyroxene*
—	—	—	—	ΔQ*	$\longrightarrow$ ⑯

⑪

$(Ab_0 + An)$		Wo	Hy	ΔQ	(resp. Hy* and ΔQ*)
-0.5		**$-Wo$**	$-Hy$	$+0.175$	*Clinopyroxene*
$(Ab_0 + An)$*		—	—	ΔQ*	$\longrightarrow$ ⑯

⑫ Calculate:
$S\ \ = 2\,X + 0.5\,Wo$ (for X see step 9)
$D\ \ = (Ab_0 + An)/S$
$Hy' = 1.15\,(Wo - X)$ and $Hy'' = Hy - Hy'$
$Pl'' = 2.5\,D\cdot X$ and $Pl' = (Ab_0 + An) - Pl''$
$Q'' = 0.35\,Pl''$ and $Q' = 0.35\,Pl'$

$(Ab_0 + An)$		Wo	Hy	ΔQ	(resp. Hy* and ΔQ*)
$-Pl''$		$-X$	$-Hy''$	$+Q''$	*Hornblende*
$-Pl'$		$-Wo$*	$-Hy'$	$+Q'$	*Clinopyroxene*
—		—	—	ΔQ*	$\longrightarrow$ ⑯

⑬ Calculate: $Hy' = 1.15\,(Wo - X)$ and $Hy'' = Hy - Hy'$
$Pl'' = 2.5\,X$ and $Pl' = (Ab_0 + An) - Pl''$
$Q'' = 0.35\,Pl''$ and $Q' = 0.35\,Pl'$

$(Ab_0 + An)$		Wo	Hy	ΔQ	(resp. Hy* and ΔQ*)
$-Pl''$		$-X$	$-Hy''$	$+Q''$	*Hornblende*
$-Pl'$		$-Wo$*	$-Hy'$	$+Q'$	*Clinopyroxene*
$(Ab_0 + An)$*		—	—	ΔQ*	$\longrightarrow$ ⑯

⑭ Claculate: $Hy'' = 2.4\,Wo + 0.6\,(Ab_0 + An)$
$Q'' = 0.35\,(Ab_0 + An)$

Ab_0	An	Wo	Hy	ΔQ	(resp. Hy* and ΔQ*)
$-Ab_0$	$-An$	$-Wo$	$-Hy''$	$+Q''$	*Hornblende*
—	—	—	Hy*	ΔQ*	$\longrightarrow$ ⑯

⑮   Calculate: $Hy'' = 3.9$ Wo and Q
                  $Pl'' = 2.5$ Wo
                  $Q'' = 0.35$ Pl''

$(Ab_0 + An)$	Wo	Hy	ΔQ	(resp. Hy* and ΔQ*)
$- Pl''$	$-$ Wo	$- Hy''$	$+ Q''$	*Hornblende*
$(Ab_0 + An)$*	—	Hy*	ΔQ*	⟶ ⑯

⑯   Distinguish two and three alternatives:
    a) $Sil^{(*)} = nil$ ⟶ ⑳
    b) $Sil^{(*)} > 0$
       ') $\Delta Q^* > 0$ ⟶ ⑰
       ") $0 > \Delta Q^* > - 0.67\ Sil^{(*)}$ ⟶ ⑱
       ''') $\Delta Q^* \leqq - 0.67\ Sil^{(*)}$ ⟶ ⑲

⑰ $\Delta Q^* = Quartz$
    $Hy^* = Orthopyroxene$ $\Big\}$ (occasionally)
    $Sil^{(*)} = Kyanite$ ⟶ ⑳

⑱

$Sil^{(*)}$	Hy*	ΔQ*	
-1.5	-1.0	$+ \Delta Q^*$	*Spinel*
$- Sil^{**}$			*Kyanite*
	$- Hy^{**}$		*Orthopyroxene*
—	—	—	⟶ ㉕

⑲

$Sil^{(*)}$	Hy*	ΔQ*	
$- Sil^{(*)}$	-0.67	$+0.67$	*Spinel*
—	Hy^{**}	ΔQ^{**}	⟶ ⑳

⑳   Distinguish four alternatives:
    a) $\Delta Q^{*(*)} > 0.07\ Hy^{*(*)}$ ⟶ ㉑
    b) $0 \leqq \Delta Q^{*(*)} \leqq 0.07\ Hy^{*(*)}$ ⟶ ㉒
    c) $0 > \Delta Q^{*(*)} > - 0.25\ Hy^{*(*)}$ ⟶ ㉓
    d) $\Delta Q^{*(*)} \leqq - 0.25\ Hy^{*(*)}$ ⟶ ㉔

㉑

Hy**	ΔQ**	
$- Hy^{**}$	-0.07	*Cummingtonite*
	$- \Delta Q^{***}$	*Quartz*
—	—	⟶ ㉕

(22)
Hy**	ΔQ**	
-14.0	$-\Delta Q$**	*Cummingtonite*
$-Hy$*		*Orthopyroxene*
—	—	⟶ (25)

(23)
Hy**	ΔQ**	
-4.0	$+\Delta Q$**	*Olivine*
$-Hy$*		*Orthopyroxene*
—	—	⟶ (25)

(24)
Hy**	ΔQ**	
$-Hy$**	$+0.25$	*Olivine*
—	ΔQ***	⟶ (25)

(25) Distinguish two alternatives:

    a) remainder: Ab_0^* [resp. $(Ab_0 + An)^*$] $\pm \Delta Q$***   ⟶ (26)

    b) remainder: ΔQ*** only   ⟶ (30)

(26) Distinguish three alternatives:

    a) ΔQ*** (resp. ΔQ**) = nil   ⟶ (27)

    b) $0 > \Delta Q$*** $> -0.4\ Ab_0^*$ ⟶ (28)

    c) ΔQ*** $\leqq -0.4\ Ab_0^*$ ⟶ (29)

(27) Ab_0^* [resp. $(Ab_0 + An)^*$] = *Albite* (resp. *Plagioclase*)

(28)
Ab_0^*	ΔQ***	
-2.5	$+\Delta Q$***	*Nepheline*
$-Ab_0$*		*Albite* (resp. *Plagioclase*)
—	—	

(29)
Ab_0^*	ΔQ***	
$-Ab_0^*$	$+0.4$	*Nepheline*
—	ΔQ_r	⟶ (30)

(30) ΔQ***, resp. ΔQ_r will be quoted as such (see Key A, step 18).

Notes: In step 25 $Ab_0^* = (Ab_0 + An)^*$, because in most instances practically all An has entered hornblende and/or clinopyroxene.
Before calculating the volume percents of the minerals (according to Key 6), add occasional riebeckite and/or cummingtonite to hornblende to form *Amphibole.*
Then enter into Key 6 (p. 131).

Key C: *Probable Mineral Assemblages of Ultramafic Rocks in the Eclogite Facies*

(1) Distinguish two alternatives:

a) the saturated norm contains *An* ⟶ (2)

b) the saturated norm contains *Ac* ⟶ (7)

(2) Convert the saturated norm into a new one, substituting Sil by Sil + 0.6 An, Wo by Wo + 0.4 An and An by nil.

(3) Calculate: $X = 0.53\,Wo + 0.94\,Sil - 0.47\,Hy\;(X_{lim} = 0)$

Sil	Wo	Hy	ΔQ	
−Sil	−X	−(2−X)	+0.33	*Garnet*
—	Wo*	Hy*	ΔQ*	⟶ (4)

(4) Distinguish two alternatives:

a) 2.6 Wo* > (Or + Ab) ⟶ (5)

b) 2.6 Wo* ≤ (Or + Ab) ⟶ (6)

(5) Calculate: $Q' = 0.1\,(2\,Wo* + Or + Ab)$

Or	Ab	Wo*	Hy*	ΔQ*	
−Or	−Ab	−Wo*	−1.13 Wo	+Q′	*Omphacite*
—	—	—	Hy**	ΔQ**	⟶ (8)

(6)	(Or + Ab)		Wo*	Hy*	ΔQ*	
	−2.6		**−Wo***	−1.13	+0.47	*Omphacite*
	(Or* + Ab*)		—	Hy**	ΔQ**	⟶ (8)

(7) Let be: M = (Ac + 2 Wo)   and calculate:
Or′ = 0.015 M   (upper limit = Or)
Ab′ = 0.125 M   (upper limit = Ab)
Q′ = 0.35 (Or′ + Ab′)

Or	Ab	Ac	Wo	Hy	ΔQ	
−Or′	−Ab′	−Ac	−Wo	−Wo	+Q′	*Clinopyroxene*
Or*	Ab*	—	—	Hy*	ΔQ*	⟶ (8)

⑧ Distinguish three alternatives (if step 2–6 was calculated, use ΔQ^{**} and Hy^{**} instead of ΔQ^* and Hy^*):

a) $\Delta Q^* > 0$
   $Hy^* = Orthopyroxene$
   $\Delta Q^* = Quartz$ ⟶ ⑨

b) $0 > \Delta Q^* > -0.25\ Hy^*$

Hy^*	ΔQ^*	
-4.0	$+\Delta Q^*$	*Olivine*
$-\mathbf{Hy^{**}}$		*Orthopyroxene*
—	—	⟶ ⑨

c) $\Delta Q^* \leqq -0.25\ Hy^*$

Hy^*	ΔQ^*	
$-\mathbf{Hy^*}$	$+0.25$	*Olivine*
—	ΔQ_1	⟶ ⑨

⑨ Distinguish three alternatives:

a) $\Delta Q_1 = nil$, then $(O_r^* + Ab^*) = Alkali\ feldspar$

b) $0 > \Delta Q_1 > -0.4\,(Or^* + Ab^*)$

$(Or^* + Ab^*)$	ΔQ_1	
-2.5	$-\mathbf{\Delta Q_1}$	*Nepheline* ($\pm$ *Kalsilite*)
$\mathbf{(Or^* + Ab^*)^*}$	—	*Alkali feldspar*

c) $\Delta Q_1 \leqq -0.4\,(Or + Ab)^*$

$(Or^* + Ab^*)$	ΔQ_1	
$-\mathbf{(Or^* + Ab^*)}$	$+0.4$	*Nepheline* ($\pm$ *Kalsilite*)
—	ΔQ_2	⟶ ⑩

⑩ Distinguish two alternatives:

a) $Garnet > 4\,\Delta Q_2$
   $Garnet - 4\,\Delta Q_2 = new\ Garnet$
   $2\,\Delta Q_2 = Olivine$
   $2\,\Delta Q_2 = Spinel$

b) $Garnet \leqq 4\,\Delta Q_2$
   $0.375\ Garnet = Olivine$
   $0.375\ Garnet = Spinel$
   $\Delta Q_2 - 0.25\ Garnet = \Delta Q_r$
   ΔQ_r will be quoted as such (see Key A, step 18).
Then enter into Key 6 (p. 131).

Table 21. List of symbols used in the Keys

a	$Ac/(Wo + Hy + Ac) =$ ordinate in Figs. 57 and 59
B_1	boundary of biotite field $(\Delta Q > 0)$ in Fig. 45
B_2	boundary of biotite field $(\Delta Q \leq 0)$ in Fig. 45
c	$(Wo + 0.4\ An')/Hy' =$ lime index in clinopyroxene.
D	$F'Wo/(x + y + z)$: e.g. $Dx = Or'$
F'	feldspar factor in clinopyroxene: $Fsp' = F'Wo$
f''	feldspar factor in amphiboles
Fsp^*	$Or^* + Ab^* + An^* =$ final feldspar
Fsp'	$Or' + Ab' + An' =$ feldspar entering clinopyroxene
Fsp''	$Or'' + Ab'' + An'' =$ feldspar entering amphibole
h	$Hy/(Hy + Sil) =$ abscissa in Figs. 41 and 42
i''	ilmenite factor in amphibole: $Il'' = i''Hy''$
k	$Or/(Or + Ab) =$ alkali ratio in the saturated norm of the rock
k'	$0.3\ k + 0.70 =$ alkali ratio in biotite
k^*	$Or^*/(Or^* + Ab^*) =$ alkali ratio in final feldspar
L	$20\ k - 6 =$ boundary of leucite field in Fig. 47
M	$(Wo + Hy)$ resp. $(Ac + Wo + Hy) =$ saturated femic compounds
m	En/Hy
n	Hy'/Wo ratio in clinopyroxene
$Ox°$	Fe^{3+}/Fe total $=$ degree of oxidation
p	ilmenite factor in clinopyroxene: $Il' = p \cdot Il$
q	$\Delta Q/(Hy + Sil) =$ ordinate in Figs. 41 and 42
q''	silica factor in amphibole in Figs. 58 and 60
q_N	silica factor in nepheline in Fig. 46
q_L	silica factor in leucite in Fig. 48
Q'	silica deficiency in clinopyroxene
Q''	silica deficiency in amphibole
ΔQ_1	silica deficiency left over after formation of femic silicates
ΔQ_m	silica deficiency left over after formation of feldspathoids
ΔQ_r	silica deficiency left over after formation of melilite, etc.
ΣQ_m	silica set free by complete conversion of clinopyroxene into melilite
R	$|\Delta Q_m|/\Sigma Q_m =$ melilite factor: $R \cdot Mel_0 =$ melilite
R'	$(|\Delta Q_m| - 0.25\ Lc)/\Sigma Q_m =$ melilite factor after formation of kalsilite
S	$Or + Ab + \Delta Q$ (saturated norm)
U	nepheline factor: $U \cdot |\Delta Q_1|$
v	$(Or + Ab + \Delta Q)/(Wo + Hy) =$ ordinate in Fig. 45
v'	$(Or + Ab)/(Wo + Hy) =$ ordinate in Fig. 47
w^*	$Wo/Hy =$ abscissa in Figs. 52 and 53
w''	Wo''/Hy'' in amphibole in Figs. 52, 53, 57 and 59
w''_1	idem, lower limit of amphibole field in Figs. 52, 53, 57 and 59
w''_2	idem, upper limit of amphibole field in Figs. 52, 53, 57 and 59
X (and Y)	various particular factors
x	orthoclase factor in clinopyroxene: $Or' = D \cdot x$
y	albite factor in clinopyroxene: $\quad Ab' = D \cdot y$
z	anorthite factor in clinopyroxene: $An' = D \cdot z$
$—'$	dash indicating components of clinopyroxene (or muscovite)
$—''$	double dash indicating components of amphibole
$—^*$	first remainder of a component, e.g. Or^*
$—^{**}$	second remainder of a component, e.g. Or^{**}
$—_0$	provisional amount of a mineral, e.g. Cpx_0, Mel_0, Ne_0 etc.

Table 22. Conversion of weight percents into number of atoms

Note: The decimals are read separately and added

Examples: $Al_2O_3 = 16.23$

Wt. %	Atoms
10	196.2
6	117.7
.2	3.9
.03	0.6
=	318.4

$FeO = 7.03$

Wt. %	Atoms
7	97.4
.03	0.4
=	97.8

SiO_2 mol $= 60.06$

Wt. %	Atoms	Wt. %	Atoms	Wt. %	Atoms	Wt. %	Atoms
10	166.5	1	16.7	0.1	1.7	0.01	0.2
20	333.0	2	33.3	0.2	3.3	0.02	0.3
30	499.5	3	50.0	0.3	5.0	0.03	0.5
40	666.0	4	66.6	0.4	6.7	0.04	0.7
50	832.5	5	83.3	0.5	8.3	0.05	0.8
60	999.0	6	99.9	0.6	10.0	0.06	1.0
70	1165.5	7	116.6	0.7	11.7	0.07	1.2
80	1332.0	8	133.2	0.8	13.3	0.08	1.3
		9	149.9	0.9	15.0	0.09	1.5

Al_2O_3 $1/2$ mol $= 50.97$

Wt. %	Atoms	Wt. %	Atoms	Wt. %	Atoms	Wt. %	Atoms
10	196.2	1	19.6	0.1	2.0	0.01	0.2
20	392.4	2	39.2	0.2	3.9	0.02	0.4
		3	58.9	0.3	5.9	0.03	0.6
		4	78.5	0.4	7.8	0.04	0.8
		5	98.1	0.5	9.8	0.05	1.0
		6	117.7	0.6	11.8	0.06	1.2
		7	137.3	0.7	13.7	0.07	1.4
		8	157.0	0.8	15.7	0.08	1.6
		9	176.6	0.9	17.7	0.09	1.8

Fe_2O_3 $1/2$ mol $= 79.84$

Wt. %	Atoms	Wt. %	Atoms	Wt. %	Atoms	Wt. %	Atoms
10	125.3	1	12.5	0.1	1.3	0.01	0.1
20	250.5	2	25.0	0.2	2.5	0.02	0.3
		3	37.6	0.3	3.8	0.03	0.4
		4	50.1	0.4	5.0	0.04	0.5
		5	62.6	0.5	6.3	0.05	0.6
		6	75.2	0.6·	7.5	0.06	0.8
		7	87.7	0.7	8.8	0.07	0.9
		8	100.2	0.8	10.0	0.08	1.0
		9	112.7	0.9	11.3	0.09	1.1

Wt. %	Atoms	Wt. %	Atoms	Wt. %	Atoms	Wt. %	Atoms
			FeO mol = 71.84				
10	139.2	1	13.9	0.1	1.4	0.01	0.1
20	278.4	2	27.8	0.2	2.8	0.02	0.3
30	417.6	3	41.8	0.3	4.2	0.03	0.4
		4	55.7	0.4	5.6	0.04	0.6
		5	69.6	0.5	7.0	0.05	0.7
		6	83.5	0.6	8.4	0.06	0.8
		7	97.4	0.7	9.7	0.07	1.0
		8	111.4	0.8	11.1	0.08	1.1
		9	125.3	0.9	12.5	0.09	1.3
			MnO mol = 70.93				
		1	14.1	0.1	1.4	0.01	0.1
		2	28.2	0.2	2.8	0.02	0.3
		3	42.3	0.3	4.2	0.03	0.4
				0.4	5.6	0.04	0.6
				0.5	7.0	0.05	0.7
				0.6	8.5	0.06	0.9
				0.7	9.9	0.07	1.0
				0.8	11.3	0.08	1.1
				0.9	12.7	0.09	1.3
			MgO mol = 40.32				
10	248.0	1	24.8	0.1	2.5	0.01	0.2
20	496.0	2	49.6	0.2	5.0	0.02	0.5
30	744.0	3	74.4	0.3	7.4	0.03	0.7
40	992.1	4	99.2	0.4	9.9	0.04	1.0
		5	124.0	0.5	12.4	0.05	1.2
		6	148.8	0.6	14.9	0.06	1.5
		7	173.6	0.7	17.4	0.07	1.7
		8	198.4	0.8	19.8	0.08	2.0
		9	223.2	0.9	22.3	0.09	2.2
			CaO mol = 56.08				
10	178.3	1	17.8	0.1	1.8	0.01	0.2
20	356.6	2	35.7	0.2	3.6	0.02	0.4
30	535.0	3	53.5	0.3	5.4	0.03	0.5
40	713.3	4	71.3	0.4	7.1	0.04	0.7
		5	89.2	0.5	8.9	0.05	0.9
		6	107.0	0.6	10.7	0.06	1.1
		7	124.8	0.7	12.5	0.07	1.2
		8	142.7	0.8	14.3	0.08	1.4
		9	160.5	0.9	16.0	0.09	1.6

Wt. %	Atoms		Wt. %	Atoms	Wt. %	Atoms	Wt. %	Atoms

Na_2O $\frac{1}{2}$ mol $= 30.997$

Wt. %	Atoms		Wt. %	Atoms	Wt. %	Atoms	Wt. %	Atoms
10	322.6	1	32.3	0.1	3.2	0.01	0.3
20	645.2	2	64.5	0.2	6.5	0.02	0.6
		3	96.8	0.3	9.7	0.03	1.0
		4	129.0	0.4	12.9	0.04	1.3
		5	161.3	0.5	16.1	0.05	1.6
		6	193.6	0.6	19.4	0.06	1.9
		7	225.8	0.7	22.6	0.07	2.3
		8	258.1	0.8	25.8	0.08	2.6
		9	290.4	0.9	29.0	0.09	2.9

K_2O $\frac{1}{2}$ mol $= 47.096$

Wt. %	Atoms		Wt. %	Atoms	Wt. %	Atoms	Wt. %	Atoms
10	212.3	1	21.2	0.1	2.1	0.01	0.2
20	424.7	2	42.5	0.2	4.2	0.02	0.4
30	637.0	3	63.7	0.3	6.4	0.03	0.6
		4	84.9	0.4	8.5	0.04	0.8
		5	106.2	0.5	10.6	0.05	1.1
		6	127.4	0.6	12.7	0.06	1.3
		7	148.6	0.7	14.9	0.07	1.5
		8	169.9	0.8	17.0	0.08	1.7
		9	191.1	0.9	19.1	0.09	1.9

TiO_2 mol $= 79.90$

		Wt. %	Atoms	Wt. %	Atoms	Wt. %	Atoms
1	12.5	0.1	1.3	0.01	0.1
2	25.0	0.2	2.5	0.02	0.3
3	37.5	0.3	3.8	0.03	0.4
4	50.1	0.4	5.0	0.04	0.5
5	62.6	0.5	6.3	0.05	0.6
6	75.1	0.6	7.5	0.06	0.8
7	87.6	0.7	8.8	0.07	0.9
8	100.1	0.8	10.0	0.08	1.0
9	112.6	0.9	11.3	0.09	1.1

P_2O_5 $\frac{1}{2}$ mol $= 70.978$

		Wt. %	Atoms	Wt. %	Atoms	Wt. %	Atoms
1	14.1	0.1	1.4	0.01	0.1
2	28.2	0.2	2.8	0.02	0.3
3	42.3	0.3	4.2	0.03	0.4
4	56.4	0.4	5.6	0.04	0.6
5	70.4	0.5	7.0	0.05	0.7
		0.6	8.5	0.06	0.8
		0.7	10.0	0.07	1.0
		0.8	11.3	0.08	1.1
		0.9	12.7	0.09	1.3

Wt. %	Atoms	Wt. %	Atoms	Wt. %	Atoms	Wt. %	Atoms
			CO_2	mol = 44.01			
10	227.2	1	22.7	0.1	2.3	0.01	0.2
20	454.4	2	45.4	0.2	4.5	0.02	0.5
30	681.7	3	68.2	0.3	6.8	0.03	0.7
40	908.9	4	90.9	0.4	9.1	0.04	0.9
		5	113.6	0.5	11.4	0.05	1.1
		6	136.3	0.6	13.6	0.06	1.4
		7	159.1	0.7	15.9	0.07	1.6
		8	181.8	0.8	18.2	0.08	1.8
		9	204.5	0.9	20.4	0.09	2.0
			Cl	mol = 35.457			
		1	28.2	0.1	2.8	0.01	0.3
		2	56.4	0.2	5.6	0.02	0.6
		3	84.6	0.3	8.5	0.03	0.8
		4	112.8	0.4	11.3	0.04	1.1
				0.5	14.1	0.05	1.4
				0.6	16.9	0.06	1.7
				0.7	19.7	0.07	2.0
				0.8	22.6	0.08	2.3
				0.9	25.4	0.09	2.5
			SO_3	mol = 80.06			
		1	12.5	0.1	1.2	0.01	0.1
		2	25.0	0.2	2.5	0.02	0.2
				0.3	3.7	0.03	0.4
				0.4	5.0	0.04	0.5
				0.5	6.2	0.05	0.6
				0.6	7.5	0.06	0.7
				0.7	8.7	0.07	0.9
				0.8	10.0	0.08	1.0
				0.9	11.2	0.09	1.1

Table 23. Abbreviations for mineral compounds

Aeg	aegirine	Kat	Catophorite
Aegaug	aegirine-augite	La	larnite
Ah	anhydrite	Lc	leucite
Åk	åkermanite	Ma	marialite
Aor	anorthoclase	Mc	monticellite
Ap	apatite	Mel	melilite
Arf	arfvedsonite	Mgs	magnesite
Aug	augite	Mln	melanite
Bi	biotite	Ms	muscovite
Bk	barkevikite	Mt	magnetite
Bn	breunnerite	Ne	nepheline
Bz	bronzite	Nfm	soda-ferrimelilite
Cc	calcite	Nm	soda-melilite
Cd	cordierite	Nos	nosean
Cm	corundum	Ns	sodasilite
Cos	cossyrite	Ol	olivine
Cpx	clinopyroxene	Phl	phlogopite
Cr	chromite	Pic	picotite
Di	diopside	Pig	pigeonite
Diag	diopsidic augite	Pigaug	subcalcic augite
Dol	dolomite	Plag	plagioclase (index = An %)
Fa	fayalite	Psk	perovskite
Fl	fluorite	Q	quartz
Fo	forsterite	Rb	riebeckite
Geh	gehlenite	Sodaug	sodian augite
Gr	garnet	San	sanidine
Hd	hedenbergite	Sid	siderite
Hl	halite	Sil	sillimanite
Hm	hematite	Sod	sodalite
Hn	haüyne	Sp	spinel
Ho	hornblende	Titaug	titanaugite
Hst	hastingsite	Th	thenardite
Hy	hypersthene	Titmt	titanomagnetite
Hz	hercynite	Tn	titanite (sphene)
Il	ilmenite	Usp	ulvöspinel
Krs	kaersutite	Z	zircon
Ks	kalsilite		

Table 24. Formulae and molecular weights of mineral components

Quartz	SiO_2		60.06

Feldspar

Orthoclase	$KAlSi_3O_8$		278.25
Albite	$NaAlSi_3O_8$		262.15
Anorthite	$CaAl_2Si_2O_8$		278.14

Feldspathoids

Leucite	$KAlSi_2O_6$		218.19
Nepheline	$NaAlSiO_4$		142.05
Kalsilite	$KAlSiO_4$		158.13

Pyroxenes

Enstatite	$MgSiO_3$		100.38
Ferrosilite	$FeSiO_3$		131.90
Wollastonite	$CaSiO_3$		116.14
Diopside	$CaMgSi_2O_6$		216.52
Hedenbergite	$CaFeSi_2O_6$		248.04
Jadeite	$NaAlSi_2O_6$		202.09
Acmite	$NaFe^{3+}Si_2O_6$		230.96
K-Acmite	$KFe^{3+}Si_2O_6$		247.06
Tschermak's molecule	$CaAl_2SiO_6$		218.08
Ti-Tschermak's molecule	$CaAlTiSiO_6$		239.01
Sodasilite	Na_2SiO_3		122.05

Amphiboles

		Mg	*Fe*
Cummingtonite	$(Mg,Fe)_{14}Si_{16}O_{44}(OH)_4$	1561.46	2002.74
Actinolite	$Ca_4(Mg,Fe)_{10}Si_{16}O_{44}(OH)_4$	1624.50	1939.70
Hornblende	$Ca_4(Mg,Fe)_8Al_4Si_{14}O_{44}(OH)_4$	1627.62	1879.78
Tschermakite	$Ca_4(Mg,Fe)_6Al_8Si_{12}O_{44}(OH)_4$	1630.74	1819.86
Pargasite	$Na_2Ca_2(Mg,Fe)_8Al_6Si_{12}O_{44}(OH)_4$	1559.27	1811.43
Hastingsite	$Na_2Ca_4(Mg,Fe)_{10}Al_4Si_{12}O_{44}(OH)_4$	1650.13	1965.33
Kaersutite	$Na_2Ca_4(Mg,Fe)_6Fe_2^{3+}Ti_2Al_2Si_{12}O_{44}(OH)_4$	1808.33	1997.45
Barkevikite	$Na_2Ca_4(Mg,Fe)_{10}Al_3Si_{13}O_{44}(OH)_4$	1659.22	1974.42
Edenite	$Na_2Ca_4(Mg,Fe)_{10}Al_2Si_{14}O_{44}(OH)_4$	1668.31	1983.51
Riebeckite	$Na_4(Fe,Mg)_6Fe_4^{3+}Si_{16}O_{44}(OH)_4$	1682.24	1871.36
Arfvedsonite	$Na_4Ca(Mg,Fe)_7AlFe_3^{3+}Si_{15}O_{44}(OH)_4$	1690.31	1910.35
Arfvedsonite II	$Na_6(Mg,Fe)_8Fe_2^{3+}Si_{16}O_{44}(OH)_4$	1665.19	1917.35
Catophorite	$Na_4Ca_2(Mg,Fe)_8AlFe^{3+}Si_{15}O_{44}(OH)_4$	1626.43	1878.49
Catophorite II	$Na_4Ca_2(Mg,Fe)_8Al_3Fe^{3+}Si_{14}O_{44}(OH)_4$	1668.31	1920.47

Micas

Muscovite	$KAl_3Si_3O_{10}(OH)_2$		398.20
Paragonite	$NaAl_3Si_3O_{10}(OH)_2$		382.10
Phlogopite	$KMg_3AlSi_3O_{10}(OH)_2$	(Biotite)	417.22
Annite	$KFe_3AlSi_3O_{10}(OH)_2$		511.78

Olivine

Forsterite	Mg_2SiO_4		140.70
Fayalite	Fe_2SiO_4		203.74

Table 24. (continued)

Garnet

Almandine	$Fe_3Al_2Si_3O_{12}$	497.64
Spessartine	$Mn_3Al_2Si_3O_{12}$	494.91
Pyrope	$Mg_3Al_2Si_3O_{12}$	403.08
Melanite	$Ca_3Fe^{3+}TiSi_3O_{12}$	508.16

Melilite

Åkermanite	$Ca_2MgSi_2O_7$	272.52
Fe-åkermanite	$Ca_2FeSi_2O_7$	304.04
Gehlenite	$Ca_2Al_2SiO_7$	274.17
Soda-melilite	$NaCaAlSi_2O_7$	258.17
Soda-ferrimelilite	$NaCaFe^{3+}Si_2O_7$	287.04

Spinels

Spinel	$MgAl_2O_4$	142.26
Hercynite	$FeAl_2O_4$	173.78
Magnetite	$Fe^{2+}Fe^{3+}_2O_4$	231.52
Ulvöspinel	$Fe^{2+}Fe^{3+}TiO_4$	231.58
Chromite	$FeCr_2O_4$	223.86

Varia

Cordierite	$Mg_2Al_4Si_5O_{18}$	584.80
Fe-Cordierite	$Fe_2Al_4Si_5O_{18}$	647.84
Titanite (Sphene)	$CaTiSiO_5$	196.04
Perovskite	$CaTiO_3$	135.98
Ilmenite	$FeTiO_3$	151.74
Apatite (F)	$Ca_5(PO_4)_3F$	503.74
Apatite (Cl)	$Ca_5(PO_4)_3Cl$	520.20

Table 25. Saturated norms of mineral components

Leucite	4 Lc	$= 5\,Or - 1\,Q$
Nepheline	3 Ne	$= 5\,Ab - 2\,Q$
Kalsilite	3 Ks	$= 5\,Or - 2\,Q$
Diopside	4 Di	$= 2\,Wo + 2\,En$
Hedenbergite	4 Hd	$= 2\,Wo + 2\,Fs$
Jadeite	4 Jd	$= 5\,Ab - 1\,Q$
Tschermak's molecule	4 Ts	$= 5\,An - 1\,Q$
Cummingtonite	30 Cum	$= 28\,Hy + 2\,Q$
Actinolite	30 Act	$= 8\,Wo + 20\,Hy + 2\,Q$
Hornblende	30 Ho	$= 10\,An + 4\,Wo + 16\,Hy$
Tschermakite	30 Tst	$= 20\,An + 12\,Hy - 2\,Q$
Pargasite	30 Par	$= 10\,Ab + 10\,An + 16\,Hy - 6\,Q$
Hastingsite	32 Hst	$= 10\,Ab + 5\,An + 6\,Wo + 20\,Hy - 9\,Q$
Kaersutite	32 Krs	$= 4\,Il + 10\,Ab + 5\,An + 6\,Wo + 12\,Hy - 5\,Q$
Barkevikite	32 Brk	$= 10\,Ab + 2.5\,An + 7\,Wo + 20\,Hy - 7.5\,Q$
Edenite	32 Edn	$= 10\,Ab + 8\,Wo + 20\,Hy - 6\,Q$
Riebeckite	30 Rb	$= 16\,Ac + 12\,Hy + 2\,Q$
Arfvedsonite	31 Arf	$= 5\,Ab + 12\,Ac + 2\,Wo + 14\,Hy - 2\,Q$
Arfvedsonite (II)	32 Arf	$= 8\,Ac + 6\,Ns + 16\,Hy + 2\,Q$
Catophorite	31 Kat	$= 5\,Ab + 4\,Ac + 3\,Ns + 4\,Wo + 16\,Hy - 1\,Q$
Catophorite (II)	32 Kat	$= 15\,Ab + 4\,Ac + 4\,Wo + 16\,Hy - 7\,Q$

Table 26. Equations of field boundaries

Fig. 40   *Factors for calculating standard magnetite*

$X_1 = 0.1 + (Na + K) \cdot 10^{-3}$
$X_2 = 1.2 \, [0.1 + (Na + K) \cdot 10^{-3}]$
$X_3 = 1.8 \, [0.07 + (Na + K) \cdot 10^{-3}]$

Fig. 41   *Field boundaries = straight-lines*

sillimanite – cordierite:	$q = 0.14 \, h$
corundum – cordierite:	$q = 0.907 \, h - 0.33$
corundum – spinel:	$q = -0.233 \, h - 0.33$
cordierite – spinel:	$h = 0.43$ (const.)
cordierite – hypersthene:	$q = 0.105 \, (1 - h)$
spinel – hypersthene:	$q = 0.756 \, (h - 1)$
spinel – olivine:	$q = 0.316 \, h - 0.566$

Fig. 42   *Field boundaries = straight-lines*

muscovite – biotite:	$q = -0.117 \, h - 0.33$
muscovite – cordierite:	$q = 0.907 \, h - 0.33$
cordierite – biotite:	$q = 0.595 - 1.245 \, h$

Fig. 43   *Feldspar factor F′ for clinopyroxenes*

if $\Delta Q > 0$, then: $F' = 0.2 \, (1 - \Delta Q \cdot 10^{-3})$
if $\Delta Q \leqq 0$, then: $F' = 0.2 - \Delta Q \cdot 10^{-3}$

Fig. 44   *Factor n for occasional hypersthene remainder* Hy*

if $\Delta Q > 0$ and $k \leqq 0.4$, then: $n = 1.3$
if $\Delta Q > 0$ and $k > 0.4$, then: $n = 1.2$
if $\Delta Q \leqq 0$ and $k \leqq 0.4$, then: $n = 1.3 + \Delta Q \cdot 10^{-3}$
if $\Delta Q \leqq 0$ and $k > 0.4$, then: $n = 1.2 + \Delta Q \cdot 10^{-3}$

Fig. 45   *Boundary of the biotite field* (k being > 0.30)

$B_1 = 6 - 6 \, k; \quad B_2 = 12 - 12 \, k$

Fig. 46   *Boundary q_N of the nepheline field and factor* U

$q_N = (3.6 - 4 \, k^*)/(9 - 3.31 \, k^*)$
$U = 1.7 - 0.5 \, (q^* + 0.5 \, k^*)$

Fig. 47   *Boundary of the leucite field* ($0.40 \leq k \leq 0.58$)

$L = 20 \, k - 6$

Fig. 48   *Boundary q_L of the leucite field* ($k^* > 0.4$)

$q_L = 1.32 \, k^*/(9 - 3.31 \, k^*)$, upper limit: $q_L = 0.2$

Table 26. (continued)

Fig. 49 *Feldspar graph for the volcanic facies*

Fig. 50 *Field boundaries = straight-lines*

muscovite – cordierite: $\quad q = \quad 0.907\,h - 0.33$
muscovite – biotite: $\qquad\; q = -0.117\,h - 0.33$
cordierite – biotite: $\qquad\;\; q = \quad 0.595 - 1.245\,h$

Fig. 51 *Field boundaries = straight-lines*

muscovite – garnet: $\qquad q = \quad 0.33\,(h - 1)$
muscovite – biotite: $\qquad q = -0.117\,h - 0.33$
garnet – biotite: $\qquad\;\; q = \quad 1.68 - 2.68\,h$

Fig. 52 *Field boundaries of calcic amphibole and clinopyroxene* $\Delta Q > 0$
         ($m = En/Hy$)

amphibole: $\qquad\qquad\qquad\quad w_1'' = 0.10 + 0.2\,m$
amphibole: $\qquad\qquad\qquad\quad w_2'' = 0.25 + 0.2\,m$
clinopyroxene, if $k > 0.4$: $\quad w' = 0.83$
clinopyroxene, if $k \leqq 0.4$: $\quad w' = 0.77$

Fig. 53 *Field boundaries of calcic amphibole* $\Delta Q \leqq 0$ ($i'' = $ ilmenite factor)

$w_1'' = 0.2\,;\quad w_2'' = 0.4$
if $m \leqq 0.5$, then: $\qquad\qquad i'' = 0.03$
if $m > 0.5$, then: $\qquad\qquad\; i'' = 0.03 + 0.8\,(m - 0.5)$
$\qquad\qquad\qquad\qquad\qquad$ (upper limit: $\;\; i'' = 1$)

Fig. 54 *Feldspar factor f'' and q'' of calcic amphibole*

if $\Delta Q > 0$, then: $\quad f'' = \quad 0.5 - 0.4\,\Delta Q \cdot 10^{-3}$
if $\Delta Q > 0$, then: $\quad q'' = \quad 0.4\,\Delta Q \cdot 10^{-3} - 0.2$
if $\Delta Q \leqq 0$, then: $\quad f'' = \quad 0.5 - 2.33\,\Delta Q \cdot 10^{-3}$
if $\Delta Q \leqq 0$, then: $\quad q'' = -0.2 - \Delta Q \cdot 10^{-3}$

Fig. 55 *Feldspar factor f' and w' for calcic clinopyroxenes*

if $\Delta Q > 0$, then: $\qquad\qquad\qquad f' = 0.15 - 0.2\,\Delta Q \cdot 10^{-3}$
if $\Delta Q > 0$ and $k > 0.4$, then: $\quad w' = 0.83$
if $\Delta Q > 0$ and $k \leqq 0.4$, then: $\quad w' = 0.77$
if $\Delta Q \leqq 0$, then: $\qquad\qquad\qquad f' = 0.15 - 0.9\,\Delta Q \cdot 10^{-3}$
if $\Delta Q \leqq 0$ and $k > 0.4$, then: $\quad w' = 0.83\,(1 - \Delta Q \cdot 10^{-3})$
if $\Delta Q \leqq 0$ and $k \leqq 0.4$, then: $\quad w' = 0.77\,(1 - \Delta Q \cdot 10^{-3})$

Fig. 56 *Ilmenite factor p for calcic clinopyroxenes*

if $k > 0.4$, then: $\quad p = 0.3\ Wo/An$
if $k \leqq 0.4$, then: $\quad p = 0.6\ Wo/An$

Table 26. (continued)

Fig. 57   *Field boundaries w' of alkaline pyroxenes and w" of alkaline amphiboles*
          ($\Delta Q > 0$)

$w' = 0.83 - 0.85\ a$ $a = Ac/(Ac + Hy)$
$w_1'' = 0.11 - 0.28\ a$
$w_2'' = 0.20 - 0.29\ a$

Fig. 58   *Feldspar factor f" and q" of alkaline amphiboles ($\Delta Q > 0$)*

$f'' = 0.14 - 2\ a$ $a = Ac/(Ac + Hy)$
$q'' = 1.3$ a

Fig. 59   *Field boundaries w' of alkaline pyroxenes and w" of alkaline amphiboles*
          ($\Delta Q \leqq 0$)

$w' = 0.83 - 0.85\ a$ $a = Ac/(Ac + Hy)$
$w_1'' = 0.16 - 0.4\ a$
$w_2'' = 0.32 - 0.4\ a$

Fig. 60   *Feldspar factors f" and q" of alkaline amphiboles ($\Delta Q \leqq 0$)*

$f'' = 0.82 - 1.5\ a$ $a = Ac/(Ac + Hy)$
if $a > 0.4$, then: $q'' = 0.05$
if $a < 0.4$, then: $q'' = 1.4\ a - 0.51$

Fig. 61   *Field boundary q_M of nepheline and factor* U

$U = 1.633 - 0.333\ (q^* - 0.5\ k^*)$

Fig. 62   *Feldspar graph for the plutonic facies*

Fig. 63   *Field boundaries = those of Fig. 50*

Fig. 64   *Field boundaries = straight-lines*

sillimanite – garnet: $q = -0.155\ h$
garnet – hypersthene: $q = 0.307\ (1 - h)$
garnet – olivine: $q = 0.99 - 1.24\ h$

Fig. 65   *Boundary q_C of the nepheline field and factor* U

$U = 1.73 - 0.33\ k^* - 0.58\ q^*$

Fig. 66   *Orthogonal graph of systematics.*

Limits of families $x = P/(A + P)$
                       $+y = Q/(Q + A + P)$
                       $-y = F/(F + A + P)$
if y is positive in sign, then: $x = 0 - 10 - 35 - 65 - 90 - 100\ \%$
if y is negative in sign, then: $x = 0 - 10 - 50 - 90 - 100\ \%$ (if $y > -60\ \%$)
                                     0———50———100 % (if $y \leqq -60\ \%$)

Comparison of the C.I.P.W. Norm with the Rittmann Norm[1]

Violetta Gottini[2]

C. I. P. W. Norms of Femic Minerals

The C.I.P.W. norm illustrates the chemical composition of igneous rocks in form of chemical compounds, called "normative minerals" calculated according to a fixed scheme. Most of these idealized minerals do not correspond to the natural rock-forming minerals, the composition of which is much more complicated. In order to illustrate this fact, the chemical compositions of some femic silicates are presented in Tables 1 and 2 in the form of their C.I.P.W. norms.

In the C.I.P.W. norm, clinopyroxenes are represented by the normative minerals enstatite (en), ferrosilite (fs), wollastonite (wo) and acmite (ac). Besides compounds of this type, natural clinopyroxenes always contain other normative minerals too. In extreme cases (no. 10 in Table 1) the amount of the latter may become even greater than that of the normative pyroxenes. Hence, the amount of modal clinopyroxene will always be greater than that of normative pyroxene. On the other hand, the amounts of many other normative minerals will be too small, and the composition of normative average feldspar will not correspond to the modal feldspars, thus prejudicing the systematic position of the rock.

The C.I.P.W. norms of amphiboles illustrate the great variability of the chemical composition of these minerals. Except for riebeckite, most amphiboles contain considerable amounts of occult salic normative minerals, as well as olivines and ores. In kaersutites and, particularly, in so-called basaltic hornblendes – which, by the way, never occur in true basalts – the sum of normative pyroxenes may become very small as e.g. in no. 10 of Table 2 (Cpx = 22 %).

A comparison of Table 1 with Table 2 shows clearly that the amphiboles cannot be considered as equivalents of clinopyroxenes. Hence, the C.I.P.W. norms of rocks containing amphiboles do not yield any reliable information about the modal composition.

---

1 The name "Rittmann norm" is proposed for the stable mineral assemblages calculated according to the new method.
2 Istituto Internazionale die Vulcanologia, I-Catania.

Table 1. C.I.P.W. norms of clinopyroxenes

C.I.P.W. norms Clinopyroxenes		1 Pig.	2 Aug.	3 Aeg. aug.	4 Aeg.	5 Pig.	6 Sal.	7 Aeg. aug.	8 Aug.	9 Ti-aug.	10 Ti-aug.
Q		1.96	—	—	3.46	—	—	—	—	—	—
or		0.71	0.17	1.12	1.06	1.36	0.71	1.30	—	—	—
ab	fsp	1.94	2.03	7.20	5.43	2.20	0.75	12.12	—	—	—
an		1.07	4.84	—	—	2.00	5.96	—	10.73	14.92	26.49
ne		—	—	0.68	—	—	0.69	1.05	3.07	4.86	1.65
lc		—	—	—	—	—	—	—	0.32	0.60	—
wo		7.42	31.94	—	10.66	13.76	44.35	43.07	36.48	35.56	22.69
en	di	2.91	16.73	—	9.88	6.58	33.96	22.15	27.34	26.44	17.08
fs		4.61	14.29	—	—	6.99	5.74	9.34	5.51	—	3.32
en	hy	28.64	13.64	—	—	27.66	—	—	—	—	—
fs		45.44	11.66	—	—	29.38	—	—	—	—	—
ac		—	—	88.38	62.89	—	—	0.73	—	—	—
ns		—	—	—	2.24	—	—	—	—	—	—
fo	ol	—	0.67	—	—	4.06	3.09	—	6.13	—	1.60
fa		—	0.63	—	—	4.76	0.57	—	1.36	—	0.34
cs		—	—	—	—	—	—	—	1.85	2.65	11.17
il		1.61	1.16	1.46	4.26	1.10	0.51	1.42	2.37	10.45	7.31
mt	ore	2.49	2.28	0.18	—	0.17	4.00	8.97	5.18	0.66	7.76
hm		—	—	0.78	—	—	—	—	—	3.97	—
ru		—	—	—	0.33	—	—	—	—	—	—
N		1.11	1.13	1.13	1.17	1.19	1.19	1.33	1.45	1.62	2.31
DEER et al., 2, p./no.		147/11	118/22	82/3	83/8	146/5	50/16	84/12	115/7	123/5	123/3

Table 2. C.I.P.W. norms of amphiboles

C.I.P.W. norm Amphiboles		1 Rb.	2 Ho.	3 Arfv.	4 Ho.	5 Krs.	6 Bkv.	7 Ho.	8 Hst.	9 Krs.	10 Bas. Ho.
Q		6.06	—	—	—	—	—	—	—	—	—
or ⎱		3.82	4.66	17.05	1.00	1.19	4.60	8.44	7.08	9.32	—
ab ⎰ fsp		—	8.79	4.59	16.06	—	4.15	2.27	—	1.91	—
an ⎰		—	5.14	—	10.81	8.98	11.00	18.67	18.69	22.36	27.00
ne		—	—	—	—	17.50	14.93	3.40	8.01	10.74	10.40
lc		—	—	—	—	5.69	—	—	3.66	—	5.79
wo ⎱		1.24	23.05	4.26	18.78	18.84	17.07	15.69	14.27	12.92	11.82
en ⎰ di		0.75	15.28	0.06	12.79	16.12	6.05	8.64	2.51	11.17	10.21
fs ⎰		44.59	6.10	4.76	4.52	0.21	11.44	6.47	12.90	—	—
en ⎱		—	13.15	0.07	6.58	—	—	—	—	—	—
fs ⎰ hy		—	5.25	5.43	2.32	—	—	—	—	—	—
ac		23.49	—	28.71	—	—	—	—	—	—	—
k-ac		5.52	—	—	—	—	—	—	—	—	—
ns		11.12	—	4.30	—	—	—	—	—	—	—
fo ⎱		—	3.19	0.34	14.06	11.20	16.08	13.96	2.89	11.38	18.03
fa ⎰ ol		—	1.40	28.09	5.47	0.16	12.66	11.52	16.37	—	—
cs ⎰		—	—	—	—	—	—	—	—	—	1.90
il ⎱		1.52	2.43	0.91	2.05	19.62	0.42	5.24	1.23	13.30	4.86
mt ⎰ ore		—	10.81	—	3.53	0.09	10.96	4.09	10.96	3.63	4.76
hm ⎰		—	—	—	—	—	—	—	—	2.09	4.38
N		1.13	1.58	2.12	2.17	2.83	2.88	3.19	3.25	4.10	4.48
DEER et al., 2. p./no.		338/2	278/24	367/9	274/10	322/2	329/3	279/26	290/23	322/5	317/3

V. Gottini

Table 3. C.I.P.W. norms of biotites

	1	2	3	4	5	6
or	2.29	6.83	—	—	—	—
an	1.88	6.44	7.24	—	—	—
lc	40.67	32.43	36.43	35.78	35.42	41.12
kp	—	—	0.11	0.45	3.36	0.47
ne	0.96	2.38	2.29	4.67	7.00	5.73
c	6.14	3.77	2.68	2.03	0.99	—
ac	—	—	—	—	—	2.73
fo	16.04	17.88	23.45	11.92	14.34	1.16
fa	16.01	12.35	13.60	21.98	27.19	28.26
mt	6.64	6.41	5.55	13.64	7.04	12.73
il	6.43	10.03	5.62	5.96	1.17	4.06

1. Biotite (Deer, Howie and Zussmann, 1962, **3**, p. 59, no. 6).
2. Biotite (ibid. p. 58, no. 4).
3. Biotite (ibid. p. 58, no. 1).
4. Lepidomelane (ibid. p. 59, no. 10).
5. Lepidomelane (ibid. p. 59, no. 8).
6. Biotite (ibid. p. 60, no. 12).

Table 4. C.I.P.W. norms of melilites

C.I.P.W. norms		1	2	3	4	5
an	fsp	0.18	0.88	—	0.96	—
lc }	foids	1.53	0.93	1.94	3.61	7.97
ne }		17.18	16.45	12.72	24.14	10.58
wo }	di	20.01	19.67	16.32	16.16	9.49
en }		14.87	14.84	11.92	10.17	8.20
fs }		3.19	2.83	2.87	4.99	—
ac		—	—	0.75	—	3.87
ns		—	—	1.58	—	—
fo }	ol	2.63	3.38	7.69	1.11	10.70
fa }		0.62	0.71	2.06	0.60	—
cs		36.19	38.52	41.48	33.55	42.36
il }	ore	0.23	—	0.11	—	—
mt }		2.75	1.41	—	3.79	1.29
hm }		—	—	—	—	5.54
N		2.60	2.64	2.97	3.13	4.63
Sum/cs		2.75	2.56	2.40	2.93	2.36

(melilite = N · (di + ac + ns)

Deer et al.,
1, p./no.   242/2   242/3   242/4   242/5   243/8

Biotites contain essential but occult normative leucite and olivine (Table 3). In the C.I.P.W. norms of oversaturated rocks containing biotite, normative orthoclase and/or hypersthene (en + fs) will appear instead of biotite. In most undersaturated rocks biotite will be replaced by the normative minerals orthoclase, nepheline and olivine. The C.I.P.W. norm will be very different from the mode.

The C.I.P.W. norms of melilites are characterized by great amounts of calcium orthosilicate (cs) in addition to wollastonite and enstatite (Table 4). Most norms of melilite-bearing rocks contain some cs, but there is no quantitative relation between cs and the amount of melilite. It is interesting to note that normative calcium aluminate (cal) is practically lacking in the C.I.P.W. norm of melilites, i.e. melilites of igneous origin do not contain or are very poor in the gehlenite component.

From the C.I.P.W. norms of femic silicates it is seen that they often contain considerable amounts of salic normative minerals. In the norms of the rocks, the salic compounds contained in the femic minerals will increase the amounts of the corresponding normative minerals. This effect will not be important in leucocratic rocks, but, the higher the colour index, the more will the C.I.P.W. norm differ from the mode and from the Rittmann norm in which the real composition of the femic minerals is taken into account. This difference will be reflected in the composition of the "average feldspar" and needs some comments.

Average Feldspars

The C.I.P.W. norm yields only an average feldspar, but no direct information about the ratio of alkali feldspar and plagioclase. As this ratio is fundamental for the systematics of igneous rocks, some authors have been led to identify the alkaline feldspar with the normative orthoclase, and the plagioclase with the sum of normative albite and anorthite. This subdivision of the average feldspar is approximately correct only if the percentage x of the normative orthoclase lies between 15 and 20, or if the percentage y of normative albite is nil or nearly so. In all other cases the results of this subdivision will be wrong, because in reality the relative amount of alkali feldspar depends on both x and y, as illustrated in Fig. 32 (p. 65).

Entering with the value of x into the graph of Fig. 1 and moving across to the intersection with the curve representing the appropriate value of y, the relative amount of alkaline feldspar can be read on the abscissa. The graph, represented in Fig. 2, serves the same purpose. It is analogous to that of Fig. 49 with the difference, however, that the coordinates are Ab_y (ordinate) and An_z (abscissa) instead of x and z

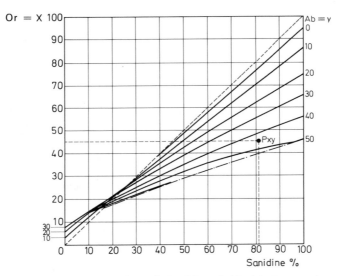

Fig. 1. Sanidine as a function of Or (x) and Ab (y). Enter x into the graph till to the intercept with y (Ab), and read sanidine % on the abscissa. The examples show that only exceptionally sanidine % $= y$; generally, sanidine % $\geqq y$ (see examples) $-$ $-$ $-$ san. % $= x$; $-$ $\cdot$ $-$ $\cdot$ $-$ limit of sanidine. Examples: 1) $Or_{45}Ab_{45}An_{10} = 81\%$ San. 2) $Or_{30}Ab_{40}An_{30} = 40\%$ San. 3) $Or_{54}Ab_{21}An_{25} = 68\%$ San. 4) $Or_{25}Ab_{20}An_{55} = 25\%$ San. 5) $Or_{12}Ab_{30}An_{58} = 6\%$ San

(an). Both graphs yield the quantities of alkaline feldspars co-existing with plagioclase in volcanic rocks. However, stress must be laid on the fact, that these nomograms can be applied to the average feldspar of the C.I.P.W. norm only, if the amounts of normative wollastonite and of nepheline are very small. Otherwise, considerable amounts of feldspar compounds will enter the femic silicates, causing discrepancies between the normative and the modal feldspars. To illustrate such differences, in Fig. 3 the average feldspars resulting from the calculation of the C.I.P.W. norm and those obtained by the Rittmann norm for the same rocks are plotted on Or-Ab-An triangles.

The effect on the average feldspar composition due to the presence of augites, biotites and/or feldspathoids can be summarized as follows:

Modal leucite very often has no counterpart (lc) in the C.I.P.W. norm, because, according to the rules of calculation, normative nepheline (ne) must be formed first in order to compensate the silica deficiency. In the case of strongly undersaturated rocks, all albite may be used up to form nepheline and, only if a deficiency of silica is still remaining, leucite will

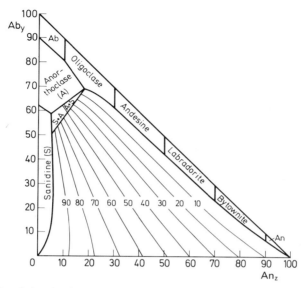

Fig. 2. Graph for the determination of the average feldspar of the C.I.P.W. norm in leucocratic volcanic rocks. *Note:* This diagram cannot be used for mesotype and melanocratic rocks, especially, if they are undersaturated!

be calculated. The remaining average feldspar will contain only Or and An, but no longer any Ab. This is a composition which is never realized in nature. Such impossible compositions will never result in the Rittmann norm, because the scheme of calculation, based on statistics, leads to the formation of leucite if the chemical composition of the rock asks for it. The average feldspar left over will be different from that yielded by the C.I.P.W. calculation. With respect to this latter, it contains less Or and

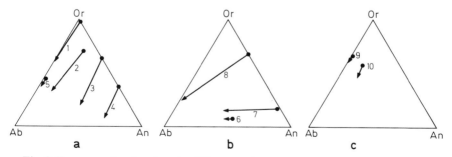

Fig. 3. Ten examples showing the difference of the average feldspar in the C.I.P.W. (full circles) and the Rittmann norm (arrow-heads) in the Or-Ab-An triangle

considerable amounts of Ab (Fig. 3a). The presence of modal biotite has a similar effect to that of leucite (Fig. 3b).

Modal nepheline always contains some potassium in contrast to the normative nepheline (ne) of the C.I.P.W. norm. Consequently, the average feldspar of the Rittmann norm will contain more albite (y) and less orthoclase (x) than that of the C.I.P.W. norm.

Calcium-rich modal clinopyroxenes, particularly titanaugites, produce great changes in the composition of the average feldspar, because they contain much occult normative anorthite. Consequently, the anorthite percentage (z) of the average feldspar yielded by the Rittmann norm will be much smaller than that resulting from the C.I.P.W. norm (Fig. 3c).

The occurrence of modal hornblendes reduces the amount of total feldspar, but the composition of the average feldspar remains practically unchanged.

C.I.P.W. Norm and Degree of Oxidation

It is important to note that, in the C.I.P.W. norm, the amount of magnetite and of occasional hematite depends upon the degree of oxidation of the analysed rock. In strongly oxidized specimens the total iron content may enter these ores and consequently, no iron silicates (fs, fa) will appear in the norm. It is evident, that such a norm does not correspond to the mode, because iron-free femic silicates do not exist. Another consequence of using up all iron to form ores is an exaggerated increase of available silica which may appear as quartz (Q) or cause an undue increase of feldspars or pyroxenes at the expense of feldspathoids or olivine. Table 5 and Fig. 4 illustrate these relations by the example of a so-called basalt of Masaya volcano/Nicaragua, Central America (R. R. Coats, 1968, p. 706).

The analysed specimen has evidently been oxidized by secondary processes, because it is quite impossible that iron is completely lacking in the pyroxenes as suggested by the C.I.P.W. norm (column A, Table 5). If the same rock had been only slightly oxidized, the C.I.P.W. norm would have been quite different, as shown in column B. No quartz, but some olivine, much more pyroxene and, naturally, much less magnetite and no hematite would have resulted. Comparing the two C.I.P.W. norms A and B, one would hardly realize at a first glance that they represent the same rock, differing only in the degree of oxidation. According to the widely accepted use of distinguishing basalts from andesites on the basis of the colour index, the strongly oxidized specimen would be an andesite, whereas the slightly oxidized one of the same lava should be called a typical basalt. Column C represents the results of the Rittmann norm.

Table 5. C.I.P.W. norms (columns A, B) for different degrees of oxidation and comparison with the Rittmann norm (column C). Example: "Basalt" of Masaya volcano/Nicaragua, Central America (R. R. Coats, 1968, p. 706)

		A	B		C		
Ox^0		0.673	0.125		0.232		
Q		5.63	—		Quartz	0.9	$Q = 1.4$
or		7.02	7.02				
ab		16.83	16.83		Sanidine	4.5	$A = 7.1$
an		34.74	34.74		Labradorite	58.2	$P = 91.5$
wo	di	10.14	10.14				
en	di	8.76	4.78				
fs	di	—	5.23		Pigeonite	33.6	
en	hy	5.63	6.82				
fs	hy	—	7.45				
fo	ol	—	1.95				
fa	ol	—	2.35				
il		0.83	0.83				
mt		10.69	2.17		Ti-magnetite	2.8	
hm		0.71	—				
CI		36.76	41.72		36.4		
an/il		41.8			$\tau = 50.1$		

Average feldspar: $or_{12.0}$ $ab_{28.7}$ $an_{59.3}$ $Or_{12.2}$ $Ab_{30.6}$ $An_{57.2}$

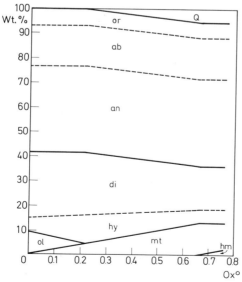

Fig. 4. C.I.P.W. norms for different degrees of oxidation. Example: "Basalt" of Masaya volcano/Nicaragua, Central America (R. R. Coats, 1968, p. 706)

Colour Index

To avoid such misleading results, the degree of oxidation, and thereby the amount of magnetite, has been standardized in the Rittmann norm according to a rule based on statistics (p. 24). This brings about definite values of the colour index in the Rittmann norm instead of the variable ones in the C.I.P.W. norm which may create ambiguities such as in the above mentioned example.

Therefore, no fixed relation can exist between the colour index yielded by the two types of norms. Even admitting equal degrees of oxidation for both, there are differences due to the following facts:

The C.I.P.W. norm yields a colour index expressed in weight percent, whereas the Rittmann norm uses volume percent. This causes a lowering of the colour index of the Rittmann norm. On the other hand, the clinopyroxenes will be more abundant in the Rittmann norm because of their content of salic components, olivine and ores, as illustrated in Table 1 on page 206. On the average, the colour index of the Rittmann norm appears slightly smaller, thus corresponding better to the volume of dark modal minerals observable in thin slides. To illustrate these facts, in Fig. 5 the colour index of the Rittmann norm is plotted against the colour index of the C.I.P.W. norm.

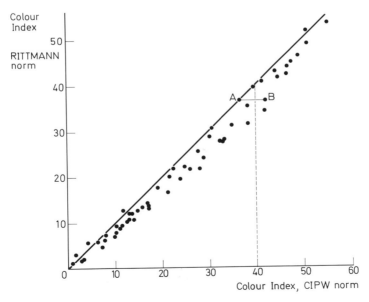

Fig. 5. Difference of the colour index in the C.I.P.W. and Rittmann norm

Conclusions

In the foregoing, the differences between the C.I.P.W. norm and the Rittmann norm and their causes have been reviewed. It must be emphasized that the Rittmann norm offers no alternative to the C.I.P.W. norm. Both types of norms are fundamentally different from each other and are based on different principles. They are not antithetical but complementary. For the discussion of purely petrochemical problems, the C.I.P.W. norm is very useful, whereas the Rittmann norm serves for petrographical purposes, such as the correct denomination and systematic position of volcanic rocks which mostly do not permit a complete determination of their modes. Furthermore, the Rittmann norm yields valuable information about the heteromorphic equivalents of igneous rocks crystallized under volcanic, subvolcanic or plutonic conditions. Its advantage lies in the fact that it makes it possible to express chemical analyses in terms used by petrographers and geologists.

ALGOL Program for the Computation of the Rittmann Norm

Wolfgang Hewers[1] and Rudolf Stengelin[2]

Purpose of the Programs

The manual computation of the Rittmann norms is rather time-consuming; the whole algorithm was therefore programmed for electronic computers. The purpose of this program is the computation of the normative mineral association of a volcanic or plutonic rock on the basis of the Rittmann norms and the output of these results.

A few additional program statements allow to store the obtained data together with the identification numbers and the chemical analyses on magnetic tape to constitute a complete data file of processed analyses. With suitable programs statistical investigations may be carried out using this file as input: mean values, variances, etc. can be computed according to certain characteristics of the analyses. Appropriate selection of the analyses required for such a statistical calculation is facilitated by adequate definition of the identification numbers (see below).

Source Language

The programming of the Rittmann norms and the testing of the programs, etc. was carried out at the "Zentrum für Datenverarbeitung der Universität Tübingen", where a CD 3300 computer system is installed. The programming language used was CD-ALGOL, a variant of ALGOL 60 which closely conforms to the definition of the international algorithmic language ALGOL[3] and the input-output procedures provided as additions thereto[4]. In the program lists the state-

1 Zentrum für Datenverarbeitung der Universität, D-74 Tübingen, Köllestraße 1.

2 Mineralogisch-Petrographisches Institut der Universität, D-74 Tübingen, Wilhelmstraße 56.

3 "Revised Report on the Algorithmic Language ALGOL 60", Communications of the ACM, **6**, pp. 1—17, 1963.

4 "A Proposal for Input-Output Conventions in ALGOL 60", Communications of the ACM, **7**, May 1964.

ments are not presented in computer notation but in a form that corresponds as closely as possible to the ALGOL report. There are a few non-essential differences from the official ALGOL 60 language (e.g. "∗" instead of " × ").

Inclusion of the Nomograms

To compute the Rittmann norms, it was necessary to include into the programs the values for the content of sanidine or orthoclase, respectively, taken from the nomograms in Figs. 49 and 62 (pp. 104 and 121, this book). The nomogram in Fig. 49 can be divided into two fields: if the point which is defined by the An or Or content falls into one of these fields, then the percentage of sanidine is determined; if it lies in the other field, only anorthoclase occurs.

In order to find out in which field the point is situated, the field boundaries, which are represented by segments of curves, are approximated by polynoms up to the sixth degree. The necessary procedures are defined at the beginning of the program. A network of points is superimposed on both the sanidine field (Fig. 49) and the orthoclase field (Fig. 62). Each point represents a certain value (percent sanidine or orthoclase) calculated from the original nomogram. There is a singular point in the sanidine field in the neighbourhood of which the sanidine values change rather rapidly. This domain is defined as follows:

$$4 \leq An \leq 10, 22 \leq Or \leq 38 .$$

In this area the points of the network are spaced at intervals of 0.5 units, measured along the axes; elsewhere they are spaced at intervals of 2 units.

The values of sanidine or orthoclase are then computed by linear interpolation between the network points. In order to carry out this interpolation with the nomogram of Fig. 49 at the field boundaries, it was necessary to add some points with fictitious values lying near but outside the sanidine field.

The values at the network points are punched on data cards and are transferred at the beginning of the program into two-dimensional arrays A, B.

Input Data

The data card file thus begins with the cards punched for the values of the nomograms; their order of sequence should not be changed. Then come the data cards proper with the values of the chemical analyses, structured according to the following list:

Columns   1 — 18 Identification number (for detailed subdivision, see the example on pp. 220 — 221

            19 — 22 Weight percent of SiO_2
            23 — 26 Weight percent of Al_2O_3
            27 — 30 Weight percent of Fe_2O_3
            31 — 34 Weight percent of FeO
            35 — 37 Weight percent of MnO
            38 — 41 Weight percent of MgO
            42 — 45 Weight percent of CaO
            46 — 49 Weight percent of Na_2O
            50 — 53 Weight percent of K_2O
            54 — 57 Weight percent of TiO_2
            58 — 60 Weight percent of P_2O_5
            61 — 63 Weight percent of Cl
            64 — 67 Weight percent of CO_2
            68 — 70 Weight percent of SO_3
            71 — 73 Weight percent of ZrO_2
            74 — 77 Weight percent of H_2O

The weight percentages are punched into the card fields without the use of a decimal point. The unit of the lowest decimal digit is equivalent to 0.01 percent. Leading blanks are interpreted as zeros.

The results can be printed in the following form:

Line 1   Identification number.
        2 — 3   Data of the chemical analysis, expressed in weight percent.
        4 — 6   Saturated norm.
        7   Values of Or, Ab and An which serve to determine the normative anorthite content with the aid of Fig. 49 or 62.
        8 — 9   Leucocratic minerals.
    10 — 15   Melanocratic minerals.
    16   Q, A, P, F and CI values for plotting the calculated rock analysis in the Streckeisen double-triangle.
    17 — 18   Additional values (see pp. 221 — 223).

If TiO_2 is not determined in the chemical analysis, no value will be printed for τ. Similarly, σ is not printed, if SiO_2 amounts to 43 weight percent.

Storage capacity required
32 K 24-bit words with the CD 3300

Computing time required
For each analysis, approx. 1.5 sec.

The Meaning of the Identification Number

For rapid identification of the geographic and stratigraphic position of a computed magmatic rock, the below defined alphanumeric system was developed. The identification number according to this system presents the following advantages: it shows the continent or oceanic area, further the country or island, from which the magmatic rock is derived and whether the rock is volcanic, plutonic or metamorphic. It includes the author's name and the date of the publication of the analysis. Finally, it identifies the material as glassy, pumiceous, ignimbritic, etc.

This identification number is subdivided as follows:

Column	1	Large areas (continents, subcontinents, oceanic areas, etc.) (see list on p. 221).
	2 – 3	Countries, islands, oceanic regions and planets.
	4	Stratigraphic position.
	5 – 7	Further geographical or petrographical subdivision.
	8 – 10	Serial number.
	11	Kind of rock (V = volcanic rock, P = plutonic rock, M = metamorphic rock).
	12 – 15	Abbreviation of the author's name or of the name of the author of a compilation of analyses.
	16 – 17	Abbreviation of the date of publication.
	18	Distinctive petrographical details (E = eclogite, G = glass, P = pumice, H = hyaloclastite, I = nonglassy ignimbrite, X = glassy ignimbrite).

Examples of identification numbers:

1IT1ROA003VGROS08–

Decoding:
1 = Europe
  IT = Italy
    1 = Cenozoic age
      RO = Roccamonfina volcanic complex
        A = Pre-caldera stage of the Roccamonfina volcanic complex
        003 = Serial analysis number
          V = Volcanic rock
            GROS = Grosse (author)
              08 = (published in) 1908
                – = (open) .

8CH2N—005VZEPI68I

Decoding:
8 = South America
  CH = Chile
    2 = Ignimbrite Formation (Late Miocene to early Pleistocene)
    N = Northern Chile
      − − = (open)
        005 = Serial analysis number
          V = Volcanic rock
            ZEPI = Zeil and Pichler
              68 = (published in) 1968
                I = Nonglassy ignimbrite .

List of codes for large areas (column 1 of the data card) :
1   Europe (excluding Iceland and Turkey).
2   Asia (including Turkey and Arabian Peninsula; excluding Japan, Kurile Islands and Kamchatka).
3   Africa (including Madagascar).
4   Australia, New Zealand and Indian Ocean (excluding Madagascar and Indonesia).
5   Atlantic Ocean (including Iceland and Arctic Region).
6   North America (including Greenland and Aleutian Islands.
7   Central America (including West Indies).
8   South America and Antarctic.
9   Pacific Ocean (including Indonesia, Japan, Kurile Islands and Kamchatka; excluding New Zealand).
0   Lunar and space material.

Output Data

The results of the program are given in a line print out of one page per analysis. An example of such an output is given in Fig. 1. In addition to the minerals for which the norm is computed, the following values are given at the end of the output:

Sigma (σ)	Serial index: $(Na_2O + K_2O)/(SiO_2 - 43)$; A. Rittmann, 1957
Tau (τ)	$(Al_2O_3 - Na_2O)/TiO_2$; V. Gottini, 1968
$\log \sigma$	
$\log \tau$	
Ox^0	Degree of oxidation
Alk	$Na_2O + K_2O$ (in weight percent)

```
NR.  1ITIROAO03VGRO5O8-     (VOLCANIC FACIES)                                   1/ 8/72

SIO2   54.75   AL2O3 20.99   FE2O3  3.07   FEO  1.08   MNO  0.10   MGO  0.78   CAO  3.43   NA2O 4.47
K2O     9.53   TIO2   0.57   P2O5   0.15   CL2  0.10   CO2  0.00   SO3  0.07   ZRO2 0.00   H2O  1.61
SUM   100.70

CC  0.0     HL  2.8     AH  1.7     TH  1.7
AP  5.6     IL 14.3     MTO  35.6
WO 45.7     HY 63.1     DELTAQ -242.4
                        AB 1011.6   OR 707.1   AN 169.7   ZIR  0.0

X (OR) 16.2    Y (AB) 64.4    Z (AN) 19.4       (FIGURE 49)


QUARTZ      0.0     SANIDINE      0.0     ANORTHOCLASE  5.3     PLAGIOCLASE 39.1     NEPHELINE 0.5
LEUCITE    43.8     KALSILITE     0.0     SODALITE      0.0     HAUYNE       1.0     NOSEAN    0.0

CLINOPYROXENE (TITANAUGITE)
OLIVINE     0.6     ORTHOPYROXENE 0.0     FAYALITE      4.6     MELILITE      0.0
MAGNETITE   1.4     SILLIMANITE   0.0     TITANITE      0.0     PEROFSKITE    0.0
ILMENITE    0.5     SPINEL        0.0     COSSYRITE     0.0     CALCITE       0.0
APATITE     0.3     CORDIERITE    0.0     SODASILITE    0.0     ZIRCON        0.0
BIOTITE     1.4     MUSCOVITE     0.0     CORUNDUM      0.0
                                         MELANITE      0.0

Q  0.0      A  5.8      P 42.9      F 51.3      CI  8.7

SIGMA  16.68    LOGSIGMA 1.222     OXO      0.70     ALK        14.00    ALK100  74.8
TAU    28.98    LOGTAU   1.462     FE(TOT)  3.94     FE(TOT)100 21.1     MGO100   4.2

                        LEUCO LEUCITE-RICH PHONOTEPHRITE
```

Fig. 1. Example of a computer output of the Rittmann norm (volcanic facies)

| Fe (tot) | $FeO + 0.9 \times Fe_2O_3 + MnO$ |
| | (in weight percent) |

$$\left.\begin{array}{l} A \\ F \\ M \end{array}\right\} \begin{array}{l} \text{proportion} \\ \text{of} \end{array} \left\{\begin{array}{l} \text{Alk} \\ \text{Fe (tot)} \\ \text{MgO} \end{array}\right.$$

in the sum Alk + Fe (tot) + MgO.

The complete ALGOL programs for the Rittmann norms (volcanic facies, wet plutonic and dry plutonic facies) are available in booklet form (price $7) from: Dr. R. STENGELIN, Mineralogisch-Petrographisches Institut der Universität, D-74 Tübingen, Wilhelmstraße 56.

Application of the Rittmann Norm Method to Petrological Problems

Hans Pichler[1] and Rudolf Stengelin[1]

The Problem of Classification of Volcanic Rocks

The results of the norm calculations in current use (C.I.P.W., Niggli's molecular norm system) in many cases agree tolerably well with the modal mineral composition. More often, however, large discrepancies arise between the norm and the mode. Especially with regard to the undersaturated volcanics the results of norm calculations according to the C.I.P.W. system often show great anomalies. This fact was recently acknowledged by Chayes and Yoder (1971).

Furthermore, both the Niggli and the C.I.P.W. norms yield only an average feldspar (or, ab, an) but no direct information about the *ratio of alkali feldspar and plagioclase*. It is this ratio which is of *fundamental importance for the systematization* of volcanic (and plutonic) rocks. Some authors have identified the alkaline feldspar with the normative orthoclase (or) and the plagioclase with the sum of normative albite and anorthite (ab + an). As pointed out by V. Gottini (p. 209), this subdivision of the average feldspar is approximately correct only if the percentage of the normative orthoclase (or) lies between 15 and 20, or if the percentage of normative albite (ab) is nil or nearly so. In all other cases this subdivision will give incorrect results and consequently the classification of the calculated magmatic rock will also be incorrect.

The discrepancies and errors which arise from the use of the common norm calculation systems are eliminated by the Rittmann norm method. With the aid of Rittmann's two diagrams (Figs. 49 and 62) illustrating the coexistence and composition of the feldspars in the volcanic and plutonic facies, the ratio of alkali feldspar to plagioclase can be determined. When the results of the Rittmann norm calculation are expressed in the form of volume percent, the plotted point of a volcanic rock in the QAPF double-triangle will determine its *correct* petrographic name.

Twelve chemical analyses of different volcanic rocks are set out together with the corresponding C.I.P.W. and Rittmann norms

1 Mineralogisch-Petrographisches Institut der Universität, D-74 Tübingen, Wilhelmstr. 56.

(Table 1a, b) in order to demonstrate the errors that can arise when the results of the C.I.P.W. norm calculation are used for the classification of volcanic rocks. The QAPF values to be entered in the Streckeisen double-triangle have been calculated from both the C.I.P.W. and the Rittmann norms of the twelve examples. In the case of the C.I.P.W. norms, the QAPF values were obtained by identifying the alkaline feldspar with the normative orthoclase (or) and the plagioclase with the sum ab + an. As mentioned above this identification leads in many cases to large errors in the naming of volcanic rocks (Fig. 1). The discrepancies are particularly great in the case of the alkaline feldspar-rich volcanics. The rock from Japan (No. 1 in Table 1a, b), which was modally determined as *rhyolite*, would become a dacite according to the C.I.P.W. norm. The modally determined *hawaiite* and the *alkali basalt* from the Hawaiian Islands (Nos. 5 and 7 in Table 1a, b) are classified as hawaiite and alkali basalt according to the Rittmann norms but, if calculated according to the C.I.P.W. system, are to be termed mugearite and nepheline-bearing latibasalt, and would fall into field 9 of the QAPF double-triangle. The *nepheline tephriphonolite* (No. 9 in Table 1a, b) changes its name to nepheline phonotephrite when we use the C.I.P.W. norm. Other examples are the *trachyte* from the Phlegrean Fields (No. 10) which must be called foid-bearing latite and the *nepheline- and haüyne-bearing latite* from Ecuador (No. 11) which becomes a mugearite according to the C.I.P.W. method. All the other cases display similar discrepancies, though less pronounced.

The advantage of the Rittmann norm in yielding a *real* mineral composition is obvious. Minerals like sanidine, anorthoclase, sodalite, haüyne, nosean, kalsilite, biotite, cordierite, spinel, cossyrite, melilite and others which do not exist in the C.I.P.W. norm may appear in the Rittmann norm (cf. Nos. 11 and 12 in Table 1b).

The colour index of rocks calculated according to the Rittmann method is, on the average, somewhat lower than that resulting from the C.I.P.W. calculation. One of the reasons for this lies in the fact that the C.I.P.W. norm yields a colour index expressed in weight percents, whereas the Rittmann norm uses volume percents. This causes a lowering of the colour index in the Rittmann norm. Furthermore, the content of magnetite and ilmenite is generally lower in the Rittmann norm, corresponding thus much better to the modal content of these opaque constituents. As an example of this we refer to No. 4 in Table 1a, b: the content of 6.03 percent magnetite and 6.23 ilmenite according to the C.I.P.W. norm is unreal for a mugearite. According to the Rittmann norm the sum of these two opaque minerals is 5.7. From our experience this amount corresponds much better to the modal composition of such a rock.

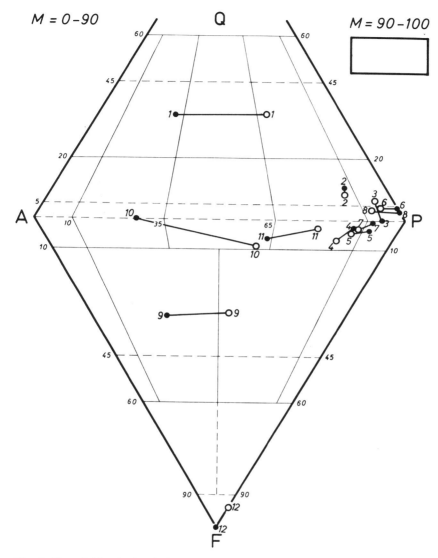

Fig. 1. Plot of 12 calculated chemical analyses (see Table 1) of various volcanic rocks in the Q–A–P–F double-triangle according to the Rittmann norm (full circles) and the C.I.P.W. norm (open circles)

H. Pichler and R. Stengelin

Table 1a. Chemical bulk analyses, C.I.P.W. and Rittmann norms of 12 selected volcanic rocks

	1	2	3	4	5	6	7	8	9	10	11	12
SiO_2	74.93	55.96	50.52	48.76	48.67	49.36	46.39	51.09	48.35	58.0	50.90	36.10
Al_2O_3	13.52	17.66	17.67	15.82	15.91	13.94	12.94	17.62	17.80	19.6	18.88	13.07
Fe_2O_3	0.76	2.25	4.57	4.10	3.24	3.03	2.05	2.64	2.93	2.1	1.82	5.28
FeO	0.72	5.73	6.98	7.53	7.13	8.53	10.16	8.42	4.65	1.9	3.20	6.68
MnO	0.08	0.17	0.18	0.17	0.15	0.16	0.17	0.21	n.d.	0.10	0.13	0.30
MgO	0.32	2.90	5.06	4.74	7.72	8.44	11.38	5.09	3.32	1.4	2.61	5.56
CaO	0.93	6.88	8.98	7.99	8.02	10.30	10.81	9.68	6.10	3.9	6.45	15.89
Na_2O	4.50	4.06	2.80	4.50	4.26	2.13	2.43	2.80	7.12	4.0	4.76	4.52
K_2O	3.39	1.58	0.58	1.58	1.28	0.38	0.93	0.76	3.92	7.2	3.45	4.96
TiO_2	0.13	1.14	1.24	3.29	2.28	2.50	2.13	1.38	2.55	0.4	0.76	3.10
P_2O_5	0.08	0.42	0.23	0.72	0.62	0.26	0.32	0.26	1.05	0.21	0.78	1.93
H_2O	0.79	0.86	1.17	—	1.06	—	0.30	0.34	0.74	0.48	3.28	1.96
											0.21	0.53
											SO_3	CO_2

C.I.P.W. *norms*

	1[a]	2	3	4	5	6	7	8	9	10	11	12[b]
Q	33.25	6.08	4.34	—	—	2.28	—	1.93	—	—	—	—
or	20.03	9.60	3.50	9.45	7.78	2.22	5.56	4.49	23.16	31.5	18.35	—
ab	38.08	37.02	25.75	30.39	30.92	17.82	17.29	23.69	18.55	33.9	34.58	—
an	4.09	25.63	34.80	18.07	20.29	27.52	21.41	33.27	5.03	14.2	21.68	0.56
ne	—	—	—	4.26	2.84	—	1.70	—	22.59	—	2.70	20.73
lc	—	—	—	—	—	—	—	—	—	8.7	—	23.11
wo ⎫ di	—			9.51	6.61	9.05	12.30	5.45	7.67			5.10
en ⎬ di	—	4.96	7.52	6.20	4.50	5.80	7.70	12.68	5.62	2.9	5.40	3.70
fs ⎭ di	—			2.64	1.58	2.64	3.83	11.39	1.32			0.92
en ⎫ hy	0.80					15.30						
fs ⎬ hy	0.63	11.86	16.96	—	—	6.60	—	—	—			—
fo ⎫ ol	—			3.92	10.36	—	14.49	—	1.85			7.14
fa ⎬ ol	—	—	—	2.04	4.08	—	7.55	—	0.48	3.2	5.20	1.84
mt	1.10	2.37	4.89	6.03	4.64	4.41	3.02	3.83	4.18	3.0	2.55	7.66
il	0.25	1.58	1.76	6.23	4.41	4.71	4.10	2.62	4.18	0.8	1.52	5.93
ap	0.19	0.90	0.48	1.68	1.34	0.67	1.01	0.60	1.90	0.5	1.68	4.37
Colour Index	2.96	21.67	31.61	38.25	37.52	49.18	54.00	36.57	27.20	10.4	16.35	53.58

[a] Including C = 0.95
[b] Including cs (dicalcium silicate) 15.82 and cc (calcite) 1.10.

Table 1b. Rittmann norms

	1	2	3	4	5	6	7	8	9	10	11	12ᵉ
Quartz	32.7	8.2	4.1	—	—	2.3	—	2.3	—	—	—	—
Sanidine	42.6	9.2	—	8.5	5.1	—	3.9	—	36.0	65.2	29.3	—
Plagioclase	19.9	64.9	67.0	59.7	60.4	52.2	45.8	66.5	15.1	24.5	51.5	23.4
Nepheline	—	—	—	2.3	2.5	—	0.3	—	24.1	—	2.2	—
Haüyne	—	—	—	—	—	—	—	—	—	—	3.3	—
Clinopyroxene	—	12.5ᵃ	24.1ᵇ	15.5ᶜ	13.7ᵃ	40.6ᵃ	27.8ᵃ	27.4ᵃ	17.5ᶜ	8.9ᵈ	5.4ᵃ	22.8ᵈ
Olivine	—	—	—	6.8	12.3	—	17.3	—	2.9	—	4.8	8.4
Biotite	2.1	0.5	—	—	—	—	—	—	—	—	—	—
Cordierite	2.3	—	—	—	—	—	—	—	—	—	—	—
Magnetite	0.2	2.6	2.9	2.7	2.5	1.9	2.4	1.9	1.9	0.9	1.0	2.4
Ilmenite	—	1.2	1.3	3.0	2.2	2.4	1.7	1.4	0.3	0.1	0.8	—
Apatite	0.2	0.9	0.5	1.5	1.3	0.6	0.7	0.6	2.2	0.4	1.6	4.5
Average plagioclase (An)	11	38	54	30	34	52	43	66	34	40	37	—
Colour Index	4.8	17.7	28.9	29.6	32.0	45.5	49.9	31.3	24.8	10.3	13.6	61.7
Sigma	1.9	2.5	1.5	6.4	5.4	1.0	3.3	1.6	22.8	8.4	8.5	−13.0
Tau	69.4	11.9	12.0	3.4	5.1	4.7	4.9	10.7	4.2	39.0	18.6	2.8
Q	34.3	10.2	5.8	—	—	4.2	—	3.3	—	—	—	—
A	44.8	11.2	—	12.1	7.5	—	7.8	—	47.9	72.4	34.0	—
P	20.9	78.6	94.2	84.7	88.8	95.8	91.5	96.7	20.1	27.6	59.6	—
F	—	—	—	3.2	3.7	—	0.7	—	32.1	—	6.4	100.0

Average clinopyroxene:
ᵃ Subcalcic augite
ᵇ Pigeonite
ᶜ Titanaugite
ᵈ Augite

ᵉ Ad 12:

Leucite	4.2
Kalsilite	10.7
Melilite	21.5
Calcite	1.4
Titanite	0.5
Perovskite	0.2

Rock names, localities and references:

1 = Rhyolite ("Aphanitic liparite", K. ONO, 1962, p. 172, No. 430)
Kōzu-shima volcano, Izu-Mariana Islands

2 = Quartz-latiandesite ("Andesitic basalt", A. R. McBIRNEY and H. WILLIAMS, 1965, Table 5, No. 2)
Concepcion volcano/Nicaragua

3 = Quartz-andesite ("Basalt", A. R. McBIRNEY and H. WILLIAMS, 1965, Table 4, No. 4)
Las Lajas Caldera/Nicaragua

4 = Olivine- and nepheline-bearing mugearite
("Average hawaiite", G. A. MacDONALD, 1960, p. 174, No. 1)
Hawaiian Islands

5 = Nepheline-bearing olivine hawaiite
("Hawaiite", G. A. MacDONALD and T. KATSURA, 1964, p. 116, No. 6)
Waianae volcano, Oahu/Hawaiian Islands

6 = Tholeiitic basalt ("Average of 181 Hawaiian tholeiitic basalts",
G. A. MacDONALD and T. KATSURA, 1964, p. 124, No. 8)
Hawaiian Islands

7 = Alkali olivine basalt ("Alkalic olivine basalt",
G. A. MacDONALD and T. KATSURA, 1964, p. 119, No. 12)
Lava of Kauai/Hawaiian Islands

8 = Andesite (CI = 31.3! "High-alumina basalt", H. KUNO, 1960, p. 125, No. 6)
Fuji volcano/Japan

9 = Nepheline tephriphonolite
("Alkali basalt", H. MÖRTEL, 1971, p. 88, No. 4340)
Vogelsberg/W. Germany

10 = Trachyte ("Phonolite", C. SAVELLI, 1967, p. 337, No. CF 43)
Astroni Crater, Phlegrean Fields, Italy

11 = Nepheline- and haüyne-bearing latite ("Andesitic tephrite",
R. J. COLONY and J. H. SINCLAIR 1928, p. 307, No. 7)
Sumaco volcano/Ecuador

12 = Leucite kalsilite melilite nephelinite
("Néphélinite mélilitique", TH. G. SAHAMA and A. MEYER, 1958, p. 75, No. 23)
Nyiragongo/E. Africa.

The Problem of Nomenclature of Igneous Rocks

The use of the Rittmann norm calculation method combined with STRECKEISEN's classification system of volcanic and plutonic rocks has the advantage that only a few rock names, which are compiled in Tables 19 and 20 (pp. 133–137), are used. The older petrographical literature, however, is overburdened with more than a thousand meaningless local rock names — compiled by W. E. TRÖGER (1935, 1939) and

A. JOHANNSEN (1931—1938) — which causes great difficulty in understanding the older literature. Therefore there is an urgent necessity to "translate" these local igneous rock names into the modern nomenclature. Such a "translation" can be realized by calculating the chemical bulk analysis associated with the respective local rock name. From the norm calculation performed according to the Rittmann method the corresponding modern rock name is easily obtained by plotting the leucocrate norm minerals into the Streckeisen double-triangle.

In Table 2 some examples of the application of the Rittmann norm method for reducing and modernizing the nomenclature of igneous rocks are given.

Table 2. Examples of rock type nomenclature according to W. E. TRÖGER (1935) and A. JOHANNSEN (1938) and their corresponding modern denomination (after the Streckeisen system; right column)

Atlantite   (TRÖGER, 1935, no. 577) = Olivine nepheline phonotephrite
Greenhalghite (TRÖGER, 1935, no. 97) = Rhyolite
Hooibergite   (TRÖGER, 1935, no. 281) = Mela amphibole monzodiorite
Khagiarite   (TRÖGER, 1935, no. 73) = Cossyrite-bearing soda quartz-trachyte
Lestiwarite   (TRÖGER, 1935, no. 170) = Aplitic soda syenite
Litchfieldite (TRÖGER, 1935, no. 415) = Leuco foid monzosyenite
Madupite   (TRÖGER, 1935, no. 645) = Haüyne melilite leucitite
Miharaite   (TRÖGER, 1935, no. 162) = Quartz-andesite
Ossipite   (TRÖGER, 1935, no. 352) = Nepheline-bearing olivine gabbro
Sancyite   (TRÖGER, 1935, no. 52) = Alkali quartz-trachyte
Sannaite   (TRÖGER, 1935, no. 498) = Mela quartz-monzonite
Tirilite   (TRÖGER, 1935, no. 81) = Granodiorite
Tokéite   (TRÖGER, 1935, no. 407) = Mela nepheline olivine tephrite
Yosemitite   (TRÖGER, 1935, no. 84) = Monzogranite

Glenmuirite   (JOHANNSEN, 4, p. 194, 1938) = Essexite
Highwoodite   (JOHANNSEN, 4, p. 41, 1938) = Foid monzosyenite
Hilairite   (JOHANNSEN, 4, p. 289, 1938) = Foid alkali (feldspar) syenite
Linosaite   (JOHANNSEN, 4, p. 68, 1938) = Nepheline-bearing olivine latibasalt
Penikkavaarite (JOHANNSEN, 4, p. 52, 1938) = Mela tonalite
Rutterite   (JOHANNSEN, 4, p. 47, 1938) = Mela alkali quartz-syenite
Sumacoite   (JOHANNSEN, 4, p. 188, 1938) = Nepheline tephriphonolite
Tannbuschite (JOHANNSEN, 4, p. 364, 1938) = Olivine tephrinephelinite
Vetrallite   (JOHANNSEN, 4, p. 174, 1938) = Leucite tephriphonolite
Westerwaldite (JOHANNSEN, 4, p. 203, 1938) = Nepheline-bearing olivine latibasalt

Comparison of Two Magmatic Regions

To compare all available chemically analysed volcanic and/or plutonic rocks of two or more magmatic regions, several different diagrammatic presentations (AFM plot, K_2O versus silica plot, etc.) can be used. In order to take in at a glance the distribution of the different rock types of such magmatic regions, the plot of the calculated rock analyses on the Streckeisen double-triangle is most suitable. This plot, however, is only possible with the aid of the calculation of the normative mineral assemblage according to Rittmann's method.

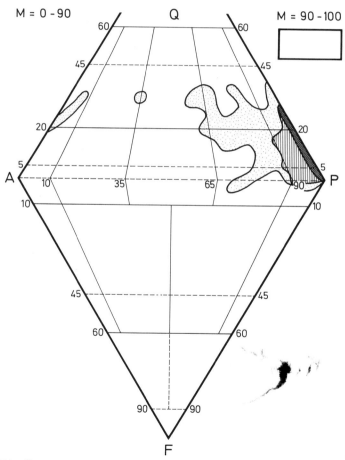

Fig. 2. Distribution areas of 170 calculated analyses of young volcanic rocks from Kamchatka and the Kurile Islands in the Streckeisen double-triangle. (Data from G. S. GORSHKOV, 1958 and V. I. VLODAVETZ and B. I. PIIP, 1959)

In Fig. 2 all available chemical analyses of the volcanic rocks of Kamchatka (G. S. Gorshkov, 1958) and the Kurile Islands (V. I. Vlodavetz and B. I. Piip, 1959) are plotted in terms of their leucocratic normative constituents. It is clearly recognizable that the bulk of the rocks is concentrated at the extreme right of the Streckeisen double-triangle, i.e. in the fields of the andesites, quartz-andesites and plagidacites. Andesites (field 10) and quartz-andesites (field 10*) represent altogether 61% of all analysed volcanics. The remainder correspond to latiandesites and quartz-latiandesites (13%), plagidacites

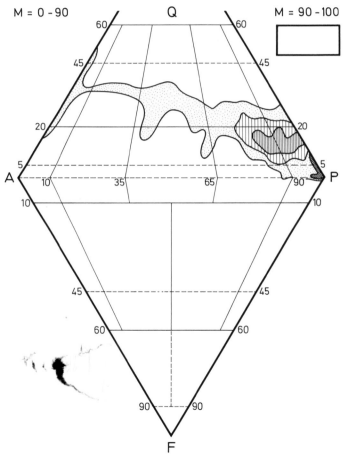

Fig. 3. Distribution areas of 342 calculated analyses of young volcanic rocks from the High Cascade Range (California, Oregon, Washington). (Data from W. S. Wise, 1969 and from literature)

(12%), dacites and rhyodacites (12.5%). With increasing content of alkali feldspar the amount of volcanic rocks decreases more and more. It is evident from Fig. 2 that both undersaturated and strongly siliceous volcanic rocks are very rare on Kamchatka and the Kurile Islands. The amount of rhyolites and alkali rhyolites reaches only 1.8%. Moreover, there exists a large gap between the rhyodacitic and alkali rhyolitic volcanics.

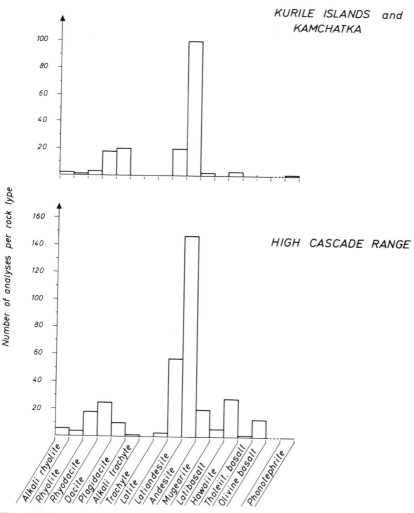

Fig. 4. Histograms of the rock types for Kamchatka and the Kurile Islands area and the High Cascade Range

The young volcanics of the High Cascade Range in the Western USA (Fig. 3) show a similar distribution in the QAPF double-triangle to those of Kamchatka and the Kurile Islands. Among the volcanics of the High Cascade Range andesites and quartz-andesites represent the main rock types, i.e. 46 %. Latiandesites and quartz-latiandesites cover 17 %, dacites and rhyodacites 12 %. Alkali rhyolites and rhyolites only amount to 2.8 %. These percentages are quite similar to those of the Kamchatka and Kurile Islands area. On the other hand, the share of the plagidacites in the High Cascade Range volcanics is remarkably lower, i.e. only 3 %. Undersaturated volcanics are also extremely rare there. Also included in the QAPF double-triangle of Fig. 3 are the basaltic and related volcanics of the Columbia River Plateau which form the base of a part of the High Cascade Range. These volcanics represent 17 % of the computed analyses.

The close similarity of the two compared areas is clearly demonstrated by the two histograms of Fig. 4. In both comagmatic regions andesitic volcanics predominate. Secondarily, there exists a smaller concentration, comprised of dacitic and related volcanics. The two concentrations are separated by a broad gap. The basalts and related rocks of the High Cascade Range area are nearly totally represented by the volcanics of the Columbia River Plateau which have no counterpart on Kamchatka and the Kurile Islands area.

Comparison of Rock Series of Various Ages within a Single Magmatic Region

Taking as an example for such studies the well-known Mt. Hood area (W. S. Wise, 1969) within the High Cascade Range, the following relations are obvious: comparing the distribution of the older (Pliocene; Fig. 5) and the younger (Pleistocene to Holocene; Fig. 6) volcanics of the Mt. Hood area with that of the whole High Cascade Range (Fig. 3) a clear conformity can be found. Figs. 5 and 6 illustrate the fact that there exists a concentration near the P (plagioclase) corner of the QAPF double-triangle due to calc-alkaline andesites and quartz-andesites which comprise 63 % of the 161 evaluated chemical analyses. A second concentration is composed of latiandesites and quartz-latiandesites (altogether 26 %). Plagidacites, dacites, rhyodacites and rhyolites form only a minor concentration (11 %).

Comparing the distribution of older and younger volcanics of Mt. Hood area in the QAPF double-triangle (Figs. 5 and 6) it can be seen that the portion of acidic volcanics diminishes strongly with decreasing age. The most acidic rock types among the younger (Pleistocene to Holocene) volcanics of the Mt. Hood area are dacites,

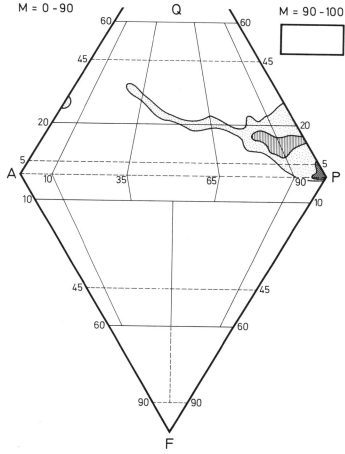

Fig. 5. Distribution areas of 109 calculated chemical analyses of Pliocene volcanics of Mt. Hood area/High Cascade Range, Oregon. (Data from W. S. WISE, 1969)

whereas among the older (Pliocene) volcanics rhyodacitic and rhyolitic types also occur. The portion of latiandesites and quartz-latiandesites likewise diminishes from 26 to 12%. On the other hand, the amount of andesites and quartz-latiandesites increases with decreasing age from 63 to 73%.

Excluding the Miocene Yakima basalts of the Mt. Hood area, which belong to the sequence of the Columbia River basalts, it was found — with the aid of the Rittmann norm method — that among neither the Pliocene nor the Pleistocene to Holocene volcanics of the Mt. Hood area do high-alumina olivine basalts occur. This stands in contrast to

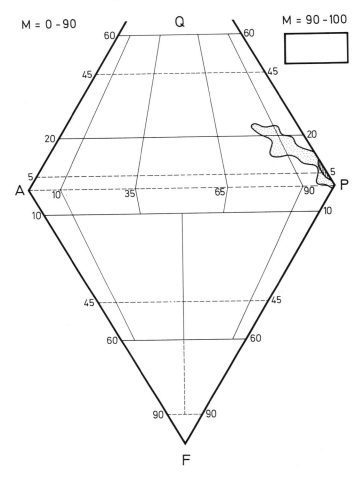

Fig. 6. Distribution areas of 52 calculated chemical analyses of Pleistocene and recent volcanics of Mt. Hood area/High Cascade Range, Oregon. (Data from W. S. WISE, 1969)

W. S. WISE (1969, p. 969). These volcanics are distinguished — according to the Rittmann norm computation — by a colour index between 4.7 and 38.1 (vol. percent). Likewise the so-called "Lookout olivine basalts" of W. S. WISE (1969, p. 994, Table 13) have a colour index lower than that of true basalts, i.e. between 23.6 and 31.8. These volcanics represent therefore not real basalts but quartz-bearing and olivine-bearing calc-alkaline andesites.

Trends of Magmatic Differentiation
Elucitated by the Rittmann Norm System

With the aid of the Rittmann norm method and its application to Streckeisen's classification scheme, trends of magmatic differentiation can often clearly be displayed. As an example for such studies the well-known north-Roman comagmatic region (Italy) has been chosen (Fig. 7). The Apenninic post-orogenic volcanics of this region belong petrogenetically to the potassic ("mediterranean") suite, characterized

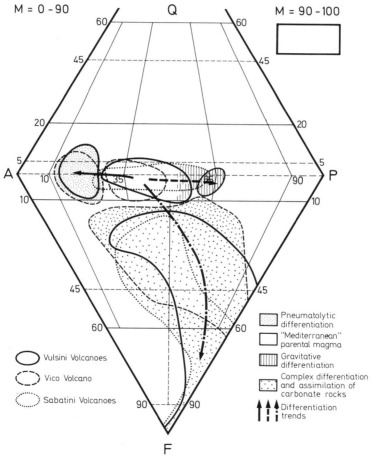

Fig. 7. Distribution of the "mediterranean" volcanic rocks of the north-Roman comagmatic region (central Italy) in terms of their magmatic evolution in the Streckeisen double-triangle. (Slightly modified after H. PICHLER, 1970, p. 31)

by the exceptionally high potassium content which is manifested in undersaturated rocks by the almost constant presence of leucite. The volcanics of the Vulsini, Vico and Sabatini areas north of Rome have been studied by several authors (H. S. WASHINGTON, 1906; C. BURRI, 1961; G. MARINELLI and M. MITTEMPERGHER, 1966; E. LOCARDI and M. MITTEMPERGHER, 1970; H. PICHLER, 1970).

For the plot of Fig. 7 all published chemical bulk analyses of the north-Roman region — about 400 — were statistically evaluated. It

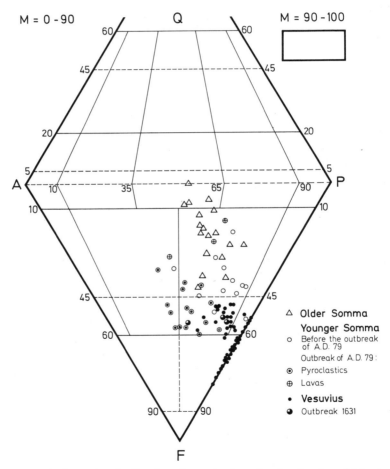

Fig. 8. Plot of all stratigrafically clear determined analysed volcanics of Monte Somma-Vesuvius (Italy) in the Streckeisen double-triangle. (After H. PICHLER, 1970, p. 166)

was found that within the three north-Roman volcanic areas the same parent magma exists, being of latitic to trachytic composition. From this parental magma three differentiation trends, very similar in all the three volcanic areas, have been developed (Fig. 7). Details of this magmatic differentiation are discussed by E. LOCARDI and M. MITTEMPERGHER (1970, p. 1089) and H. PICHLER (1970, p. 28).

Trends of Magma Evolution Displayed with the Aid of the Rittmann Norm System

The evolution of the Mt. Somma-Vesuvius magmas was clearly demonstrated by A. RITTMANN (1933) who took a progressive assimilation of dolomite into account. Because the roof of the magma chamber consists of Triassic dolomites, assimilation of these rocks must necessarily be assumed, the more so in the light of the evidence provided by the numerous endomorphically and exomorphically altered ejecta.

In the QAPF double-triangle of Fig. 8 all those normatively calculated analyses of Mt. Somma-Vesuvius are projected which are stratigraphically clearly attributable to the different eruptive periods in the history of this world-famous volcano (Older Somma, Younger Somma, Vesuvius). The oldest volcanics, i.e. those of the Older Somma period, are leucite-bearing latites and leucite phonotephrites, whereas the volcanics of the Younger Somma period are represented by stronger undersaturated leucite phonotephrites and leucite tephriphonolites. The youngest lavas and pyroclastics, those of Mount Vesuvius proper, comprise strongly subsilicic leucite phonotephrites, leucite tephrites and tephrileucitites (Fig. 8). The increase of leucite during the temporal evolution of Mt. Somma-Vesuvius caused by assimilation of dolomite in connexion with gravitative differentiation and gaseous transfer is indicated by the black arrow in Fig. 9. Starting from a parental magma of leucite-bearing latitic composition, the continual assimilation of dolomite led to a tephrileucititic magma which corresponds to the present-day Vesuvius.

Comparison of Volcanic Rocks with Holocrystalline Plutonic Ejecta

Within the volcanic sequences of the Santorini group (Aegean Sea, Greece) numerous holocrystalline inclusions and ejecta occur (S. KUSSMAUL, 1971; H. PICHLER and S. KUSSMAUL, 1971). These subvolcanic rocks can be subdivided into two broad groups distinguished by their origin:

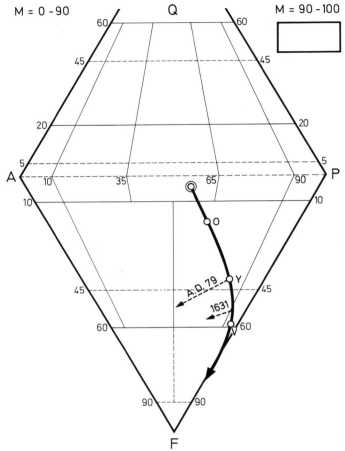

Fig. 9. Evolution and differentiation of the magma of Monte Somma-Vesuvius (Italy), plotted in the Streckeisen double-triangle. The thick arrow shows the evolution of the magma caused by assimilation of dolomite and strong gaseous transfer; the two small, broken arrows indicate the differentiations which took place before the paroxysmal outbreaks of 79 A. D. and 1631. These differentiations are superimposed upon the main magmatic evolution. Double circle = "Mediterranean" parental magma, O = Older Somma, Y = Younger Somma, V = Vesuvius. (After H. PICHLER, 1970, p. 131, modified according to A. RITTMANN, 1933, p. 29)

a) Igneous cumulates in the sense of WAGER et al. (1960), formed gravitatively by crystal accumulation in magma chambers. The plot of these orthocumulates in the STRECKEISEN double-triangle shows a concentration in the plagioclase corner. If we calculate the analyses of these orthocumulates according to the Rittmann norm, specifically in

Table 3. Rittmann norms (volcanic facies) of two orthocumulates, two holo-crystalline inclusions and of two andesitic lavas from Santorini volcano complex (Aegean Sea/Greece; after S. KUSSMAUL, 1970)

	1	2	3	4	5	6
Quartz	—	—	2.7	8.4	5.7	0.6
Sanidine	—	—	—	4.4	1.4	—
Plagioclase	73.5	77.1	67.8	59.9	64.6	68.1
An %	70	80	51	53	48	55
Nepheline	1.7	2.7	—	—	—	—
Clinopyroxene	10.4	15.8	26.3	24.3	23.8	28.4
Olivine	11.4	2.7	—	—	—	—
Magnetite	2.2	1.1	2.2	2.0	2.8	1.8
Ilmenite	0.9	0.6	0.9	0.8	1.0	1.0
Apatite	—	—	—	0.2	0.6	0.2
Colour Index	24.8	20.2	29.5	27.3	28.2	31.4

1 = Orthocumulate, inclusion or ejecta in the volcanics of Cape Akrotiri
2 = Orthocumulate, inclusion in the Niki lavas (1940—41) of Nea Kameni
3 = Holocrystalline inclusion, Cape Akrotiri
4 = Holocrystalline inclusion in the lava from 1925, Nea Kameni
5 = Andesitic lava from Cape Turlos (Skaros)
6 = Quartz-andesitic lava from Cape Alonaki

volcanic facies, significant differences are found between the norms of these rocks and those of the true andesitic volcanics which occur abundantly on Santorini. Most of these latter volcanics are quartz-normative; normative nepheline never occurs. The norms of the orthocumulates, however, calculated in volcanic facies, show in most cases a slight amount of nepheline; normative free quartz is never found (Table 3). With regard to the colour index, conspicuous differences exist between the andesites and orthocumulates. The colour index of the orthocumulates is significantly lower than that of the andesites.

b) By far the greater part of the holocrystalline ejecta and inclusions are real subvolcanic rocks, i.e. crystallized in apophyses of the magma chamber(s) under subvolcanic conditions (namely higher water-pressure). Calculating the chemical analyses of this group of subvolcanic rocks according to the Rittmann norm, specifically in volcanic facies, a close similarity to the norms of the volcanics of Santorini is found. The volcanic rocks of the Santorini group range in composition from rhyodacites to andesites, but are predominantly quartz-latiandesites. Plotting the norms (in volcanic facies) of the volcanic rocks and those of the second group of subvolcanics into the QAPF double-triangle a close similarity in the distribution of both types is seen to exist (Fig. 10).

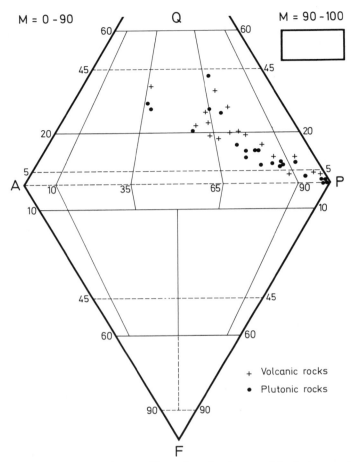

Fig. 10. Distribution of holocrystalline plutonic ejecta and inclusions and of the corresponding volcanic rocks from the Santorini volcano complex (Aegean Sea/ Greece). (After S. KUSSMAUL, 1970, p. 90)

From this fact it may be concluded that the second group of subvolcanic rocks represent the subvolcanic equivalents of the Santorini volcanics. Therefore the application of the Rittmann norm method permits one to group the holocrystalline inclusions and the ejecta either with the cumulates or with the real subvolcanic rocks.

Comparison of Pitchstones and Their Residual Glasses

In general the nomenclature of natural glasses is erroneously based on the small amount of phenocrysts occurring in these glasses. The

Table 4. Chemical analyses and Rittmann norms of 5 pitchstones and their residual glasses. Chemical analyses from I.S.E. CARMICHAEL (1960, p. 327). The values of the analyses are in weight %, those for the Rittmann norms in volume %

	1		2		3		4		5	
	a	b	a	b	a	b	a	b	a	b
SiO$_2$	66.0	69.4	69.2	69.0	69.8	72.3	64.4	67.0	71.9	73.6
Al$_2$O$_3$	15.1	14.3	12.0	10.9	12.9	12.1	12.7	12.2	11.7	11.7
Fe$_2$O$_3$	1.5	1.3	2.2	1.2	0.7	0.4	1.7	1.2	0.6	0.4
FeO	1.2	1.0	1.4	1.8	2.4	1.4	3.8	3.9	2.1	1.1
MnO	0.11	0.11	0.12	0.11	0.08	0.05	0.21	0.20	0.07	0.02
MgO	0.69	0.41	0.59	0.40	0.21	0.09	0.66	0.32	0.38	0.14
CaO	1.2	0.92	2.2	1.6	1.7	1.0	1.9	1.4	1.4	0.70
Na$_2$O	4.9	4.7	4.1	4.1	4.8	4.9	4.9	3.9	4.0	4.1
K$_2$O	6.2	5.9	1.6	1.8	3.1	3.5	3.7	4.4	3.5	4.5
TiO$_2$	0.71	0.59	0.50	0.37	0.34	0.22	0.69	0.55	0.37	0.18
P$_2$O$_5$	0.11	0.10	0.08	0.06	0.07	0.04	0.11	0.11	0.05	0.03
H$_2$O	2.1	1.6	5.9	8.6	3.8	3.9	5.2	4.4	3.8	3.3
Rittmann norms:										
Quartz	11.4	17.6	35.5	37.5	27.2	29.5	18.8	25.5	33.0	31.6
Sanidine	81.7	77.5	6.9	2.8	30.8	—	—	62.6	48.4	65.6
Anorthoclase	—	—	—	—	—	67.0	69.4	—	—	—
Plagioclase	—	—	52.0	54.8	37.1	—	—	1.4	12.2	—
			An 19	An 12	An 08			An 08	An 09	
Clinopyro-										
xene	3.7	3.1	0.7	1.3	2.8	2.7	6.6	2.8	2.5	2.3
Orthopyroxene	0.7	0.4	3.4	2.3	0.7	—	1.4	2.2	1.1	—
Biotite	1.2	0.6	—	—	—	—	2.1	3.6	1.8	—
Titanite	—	—	—	—	—	0.2	—	—	—	—
Magnetite	0.7	0.6	0.8	0.7	0.9	0.5	1.6	1.4	0.7	0.4
Ilmenite	0.2	0.1	0.6	0.4	0.3	—	—	0.3	0.2	—
Apatite	0.2	0.2	0.2	0.1	0.1	0.1	0.2	0.2	0.1	0.1
Colour Index	6.8	5.0	5.6	4.9	4.8	3.5	11.8	10.6	6.4	2.8

Key to the analyses:
a = pitchstone, b = residual glass.
1 a, b: Sgurr of Eigg/Scotland.
       Correspond to *alkali quartz-trachyte.*
2 a, b: Glass from margin of acid lava, Thingmuli/east Iceland.
       a = *Dacite,*
       b = *Leuco plagidacite.*
3 a, b: From glassy margin of dyke, Thingmuli/east Iceland.
       a = *Rhyodacite,*
       b = *Alkali rhyolite.*
4 a, b: From margin of intrusion, Bondolfur/east Iceland.
       Correspond to *mela alkali rhyolite.*
5 a, b: Glen Shurig, Arran/Scotland.
       a = *Rhyolite,*
       b = *Alkali rhyolite.*

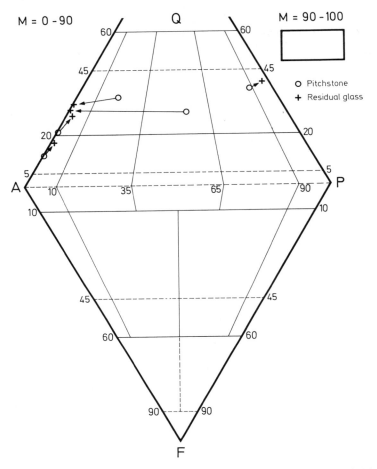

Fig. 11. Plot of 5 calculated chemical analyses of pitchstones and their residual glasses in the Streckeisen double-triangle. Chemical analyses from I. S. E. CARMICHAEL (1960, p. 327)

vitreous part which forms the bulk of such volcanics is mostly not considered in the naming of these glassy rocks. Because the phenocrysts have a more basic chemical bulk composition than the surrounding glass, the name of such rocks, which is derived only from these phenocrystic constituents, does not refer to the real chemical composition. If no phenocrysts are present, vague terms like obsidian, pitchstone, pumice, sideromelane, palagonite, etc. are used. With the aid of the Rittmann norm method it is possible to classify and denominate these types of volcanic rocks with respect to the *total* composition, i.e. both glass and phenocrysts.

It is a well-known fact that during the cooling of a volcanic melt the dark constituents are generally crystallizing first. In the course of the crystallization the silica content of the residual melt is usually increasing more and more. This fact can be clearly demonstrated by the analytical data of pitchstones and their residual glasses from Iceland and Northern Scotland, published by I. S. E. CARMICHAEL (1960, Table 6). These residual glasses, i.e. the pure glasses after separation of the phenocrysts, show a higher content of silica but lower levels in TiO_2, Fe_2O_3, MgO and CaO (cf. Table 4), caused by the crystallization of the femic minerals. The different composition of the phenocrysts and the surrounding glass can clearly been shown with the aid of the Rittmann norm, calculating the normative composition both of the pitchstones and their residual glasses. In Table 4, the normative composition of five pitchstones and their residual glasses (chemical analyses from I. S. E. CARMICHAEL 1960) are given. Fig. 11 shows the position of these glassy volcanics in the QAPF double-triangle. With exception of analyses 5a and b, the amount of normative quartz is higher in the residual glasses than in the whole (pitchstone) rocks. This difference is so strong that in three examples (analyses 2a, b; 3a, b and 5a, b) different rock names are resulting. Thus, for instance, the pitchstone from Thingmuli/Iceland (Nr. 3) must be named rhyodacite, whereas its residual glass has alkali rhyolitic composition.

References

AHRENS, W., VILLWOCK, R.: Exkursion in den Westerwald am 6. September 1964. Fortschr. Mineral. **42**, 303—320 (1964).

BARTH, T. F. W.: Zonal structure in feldspars of crystalline schists. 3rd Réunion Internat. Réactivité à l'état solid, **3**, p. 363, Madrid 1956.

BURRI, C.: Le province petrografiche postmesozoiche dell'Italia. Rend. Soc. Mineral. Ital. **17**, 3—40 (1961).

CARMICHAEL, I. S. E.: The pyroxenes and olivines from some Tertiary acid glasses. J. Petrol. **1**, 309—336 (1960).

CHAYES, F., YODER, H. S.: Some anomalies in the norms of extremely undersaturated lavas, Carnegie Institution Year Book, **70**, 205—206 (1971).

COATS, R. R.: Basaltic andesites. In: HESS, H. H. and POLDERVAART, A. (Ed.): Basalts, **2**, 689—736. New York–London: J. Wiley & Sons 1968.

COLONY, R. J., SINCLAIR, J. H.: The lavas of the volcano Sumaco, eastern Ecuador, South America. Am. J. Sci. **216**, 299—312 (1928).

CROSS, W., IDDINGS, J. P., PIRSSON, L. V., WASHINGTON, H. S.: Quantitative classification of igneous rocks, 286 pp., Chicago: Univ. Chicago Press 1903.

DEER, W. A., HOWIE, R. A., ZUSSMANN, J.: Rock-forming minerals. 5 Volumes. London: Longmans 1962/63.

DENAYER, M.-E., SCHELLINCK, F. [with contributions of COPPEZ, A.]: Recueil d'analyses des laves du fossé tectonique de l'Afrique centrale (Kivu, Rwanda, Toro-Ankole). — Ann. Musée Royal de l'Afrique centrale, Ser. In-8°, Sci. géol., no. 49, IX + 234 pp., Tervuren/Belgique 1965.

DE VECCHI, G., OMENETTO, P., SCOLARI, A.: Verifica del metodo RITTMANN-VIGHI nello studio modale dei minerali opachi di alcuni "basalti" veneti. Ric. sci. **38**, 1108—1111 (1968).

ENGELHARDT, W. v., WEISKIRCHNER, W.: Einführung zu den Exkursionen der Deutschen Mineralogischen Gesellschaft zu den Vulkanschloten der Schwäbischen Alb und in den Hegau. Fortschr. Mineral. **40**, 5—28 (1962).

GORSHKOV, G. S.: Kurile Islands. Catalogue of the Active Volcanoes of the World including Solfatara Fields, Part 7, 99 pp. Napoli: IVA 1958.

GOTTINI GRASSO, V.: The TiO_2 frequency in volcanic rocks. Geol. Rundsch., **57**, 930—935 (1968).

GOTTINI, V.: Serial character of the volcanic rocks of Pantelleria. Bull. volcanol. **33**, 818—827 (1970).

GOTTINI, V.: Some remarks on contact anatexis. Bull. volcanol. **34**, 406—413 (1971).

JOHANNSEN, A.: A descriptive petrography of the igneous rocks, **4**, XVII + 523 pp. Chicago: Univ. Chicago Press 1938.

JUNG, D.: Untersuchungen am Tholeyit von Tholey (Saar). Beitr. Mineral. Petrol. **6** (1957/59), 147—181 (1958).

JUNG, D.: Note concerning the use and significance of the term "tholeyite" (= tholeiite). Neues Jahrb. Mineral. Abh. **109**, 267—273 (1968).

KLERKX, J.: Problèmes concernant la détermination des volcanites basiques. Ann. Soc. Géol. Belgique **90**, 771—778 (1967).

KUNO, H.: High-alumina basalt. J. Petrol. **1**, 121—145 (1960).

KUSSMAUL, S.: Vulkanologie und Petrographie von Nord-Thera (Santorin-Gruppe/Ägäis). Diss. Univ. Tübingen, V + 112 S., Tübingen 1971.

LOCARDI, E., MITTEMPERGHER, M.: The Meaning of Magmatic Differentiation in Some Recent Volcanoes of Central Italy. Bull. volcanol. 33, 1089—1100 (1970).

MACDONALD, G. A.: Dissimilarity of continental and oceanic rock types. J. Petrol. 1, 172—177 (1960).

MACDONALD, G. A.: Physical properties of erupting Hawaiian magmas. Bull. Geol. Soc. Am. 74, 1071—1078 (1963).

MACDONALD, G. A., KATSURA, T.: Chemical composition of Hawaiian lavas. J. Petrol. 5, 82—133 (1964).

MARINELLI, G., MITTEMPERGHER, M.: On the Genesis of some Magmas of Typical Mediterranean (Potassic) Suite. Bull. volcanol. 29, 113—140 (1966).

MAZZUOLI, R.: Le vulcaniti di Roccastrada (Grosseto). Atti Soc. Toscana Sci. Nat. Mem., Ser. A, 74, 315—373 (1967).

MCBIRNEY, A. R., WILLIAMS, H.: Volcanic history of Nicaragua. Univ. California Publ. Geol. Sci. 55, 1—65 (1965).

MINAKAMI, T., SAKUMA, S.: Report on volcanic activities and volcanological studies concerning them in Japan during 1948—1951. Bull. volcanol., Ser. II, 14, 79—130 (1953).

MITTEMPERGHER, M.: Vulcanismo e petrogenesi nella zona di San Venanzo (Umbria). Atti Soc. Toscana Sci. Nat., Ser. A, 72, 437—479 (1965).

MÖRTEL, H.: Foide und Zeolithe (Restkristallisate) basaltischer Gesteine des Vogelsberges (Hessen). Neues Jahrb. Mineral. Abh. 115, 54—97 (1971).

NIGGLI, P.: Die quantitative mineralogische Klassifikation der Eruptivgesteine. Schweiz. Mineral. Petrogr. Mitt. 11, 296—364 (1931).

NIGGLI, P.: Über Molekularnormen zur Gesteinsberechnung. Schweiz. Mineral. Petrogr. Mitt. 16, 295—317 (1936).

NIGGLI, P.: Die Magmentypen. Schweiz. Mineral. Petrogr. Mitt. 16, 335—399 (1936).

NOCKOLDS, S. R.: Average chemical compositions of some igneous rocks. Bull. Geol. Soc. Am. 65, 1007—1032 (1954)

ONO, K.: Chemical composition of volcanic rocks in Japan. 441 pp. Tokyo: 1962.

PICHLER, H.: Italienische Vulkan-Gebiete I (Somma-Vesuv, Latium, Toscana). Sammlung geol. Führer, Bd. 51, XIII + 258 S. Stuttgart: Gebr. Borntraeger 1970.

PICHLER, H., STENGELIN, R.: Petrochemische und nomenklatorische Revision der Vulkanite des südägäischen Raumes (Griechenland). Geol. Rundsch. 57, 795—810 (1968).

POLDERVAART, A., HESS, H. H.: Pyroxenes in the crystallization of basaltic magma. J. Geol. 59, 472—489 (1951).

RITTMANN, A.: Die Zonenmethode. Ein Beitrag zur Methodik der Plagioklas-bestimmung mit Hilfe des Theodolithtisches. Schweiz. Mineral. Petrogr. Mitt. 9, 1—46 (1929).

RITTMANN, A.: Die geologisch bedingte Evolution und Differentiation des Somma-Vesuvmagmas. Zschr. Vulkanol. 15 (1933/34), 8—94 (1933).

RITTMANN, A.: Nomenclature of Volcanic Rocks. Bull. volcanol., Ser. II, 12, 75—102 (1952).

RITTMANN, A.: On the serial character of igneous rocks. Egyptian J. Geol. 1, 23—48 (1957).

RITTMANN, A.: Le cause della corrosione magmatica. Boll. Accad. Gioenia Sci. Nat. Catania, Ser. IV, 4, 534—541 (1958).

RITTMANN, A.: Vulkane und ihre Tätigkeit. 336 S. Stuttgart: Enke 1960.

RITTMANN, A.: Volcanoes and their activity, XIV + 305 pp. New York–London: J. Wiley & Sons 1962.

RITTMANN, A.: Die Bimodalität des Vulkanismus und die Herkunft der Magmen. Geol. Rundsch. 57 (1967/68), 277—295 (1967).

RITTMANN, A.: Note to the contribution by V. GOTTINI on the "Serial character of the volcanic rocks of Pantelleria". Bull. volcanol. 33, 979—981 (1970).

RITTMANN, A.: The probable origin of high-alumina basalts. Bull. volcanol. 34, 414—420, (1971).

RITTMANN, A., EL-HINNAWI, E. E.: The application of the zonal method for the distinction between low- und high-temperature plagioclase feldspars. Schweiz. Mineral. Petrogr. Mitt. 40, 41—48 (1960).

RITTMANN, A., VIGHI, L.: Fattore di correzione da apportare alle percentuali di minerali opachi determinati nelle sezioni sottili col metodo di Rosiwal. Period. Mineral. 16, 109—122 (1947).

ROMANO, R.: Sur l'origine de l'excès de sodium (ns) dans certaines laves de l'Ile de Pantelleria. Bull. volcanol. 33, 694—700 (1969).

SAHAMA, TH. G., MEYER, A.: Study of the volcano Nyiragongo. Inst. Parcs Nat. du Congo belge. Exploration du Parc National Albert. Mission étud. volcanol. Fasc. 2, 1—85 (1958).

SAVELLI, C.: The problem of rock assimilation by Somma-Vesuvius magma. Contrib. Mineral. Petrol. 16, 328—353 (1967).

SCHERILLO, A.: Le lave e le scorie dell'eruzione vesuviana del marzo 1944. Ann. Osserv. Vesuviano, Ser. V, 1, 169—183 (1949).

STRECKEISEN, A.: Die Klassifikation der Eruptivgesteine. Geol. Rundsch. 55, 478—491 (1966).

STRECKEISEN, A.: Classification and nomenclature of igneous rocks. Neues Jahrb. Mineral. Abh. 107, 144—240 (1967).

STRECKEISEN, A.: Account of classification and nomenclature of igneous rocks. Reprint 23rd Internat. Geol. Congr. Prague, 1968.

STRECKEISEN, A., ed.: Report of the Preliminary Meeting of the Subcommission on the Nomenclature and Systematics of Igneous Rocks of I.U.G.S., held at Berne, April 11—14, 1972. Contr. no. 26 of the Subcommission (including Reports of the Working Group on Basic and Ultramafic Rocks [Chairman: F. ROST] and of the Working Group on Alkaline and Feldspathoidal Rocks [Chairman: H. SØRENSEN]), Berne 1972.

TRÖGER, W. E.: Spezielle Petrographie der Eruptivgesteine. 360 S. Berlin: Deutsche Mineral. Ges. 1935.

TRÖGER, W. E.: Eruptivgesteinsnamen. Fortschr. Mineral. 23, 41—90 (1939).

TRÖGER, W. E.: [Edited by BAMBAUER, H. U., TABORSZKY, F., TROCHIM, H. D.] Optische Bestimmung der gesteinsbildenden Minerale. Teil 2 (Textband). 822 S. Stuttgart: E. Schweizerbart 1969.

VLODAVETZ, V. I., PIIP, B. I.: Kamchatka and continental areas of Asia. Catalogue of the Active Volcanoes of the World including Solfatara Fields, Part 8, 110 pp. Napoli: IVA 1959.

WAGER, L. R., BROWN, G. M., WADSWORTH, W. J.: Types of Igneous Cumulates. J. Petrol. 1, 73—85 (1960).

WASHINGTON, H. S.: The Roman Comagmatic Region. Carnegie Inst. Publ. No. 57, VI + 199 pp. (1906).

WISE, W. S.: Geology and petrology of the Mt. Hood area: A study of High Cascade volcanism. Geol. Soc. Am. Bull. 80, 969—1006 (1969).

WOLFF, F. VON: Die Prinzipien einer quantitativen Klassifikation der Eruptivgesteine, insbesondere der jungen Ergußgesteine. Geol. Rundsch. 13, 9—17 (1922).

YODER, H. S., TILLEY, C. E.: Origin of basalt magmas: an experimental study of natural and synthetic rock systems. J. Petrol. 3, 342—532 (1962).

Subject Index

Minerals, Rocks and Inorganic Materials

G. M. BROWN and D. H. LINDSLEY: Synthesis and Stability of Pyroxenes
J. J. FAWCETT: Synthesis and Stability of Chlorite and Serpentine
D. H. LINDSLEY and S. HAGGERTY: Synthesis and Stability of Iron-Titanium Oxides

Subseries "Isotopes in Geology":

D. C. GREY: Radiocarbon in the Earth System
H. P. TAYLOR: Oxygen and Hydrogen Isotopes in Petrology

Subseries "Crystal Chemistry of Non-Metallic Materials"

R. E. NEWNHAM and R. ROY: Crystal Chemistry Principles
R. ROY and O. MULLER: The Major Binary Structural Families